INFRARED SPECTROSCOPY:
FUNDAMENTALS AND
APPLICATIONS

Analytical Techniques in the Sciences (AnTS)

Series Editor: David J. Ando, Consultant, Dartford, Kent, UK

A series of open learning/distance learning books which covers all of the major analytical techniques and their application in the most important areas of physical, life and materials sciences.

Titles Available in the Series

Analytical Instrumentation: Performance Characteristics and Quality
Graham Currell, University of the West of England, Bristol, UK

Fundamentals of Electroanalytical Chemistry
Paul M. S. Monk, Manchester Metropolitan University, Manchester, UK

Introduction to Environmental Analysis
Roger N. Reeve, University of Sunderland, Sunderland, UK

Polymer Analysis
Barbara H. Stuart, University of Technology, Sydney, Australia

Chemical Sensors and Biosensors
Brian R. Eggins, University of Ulster at Jordanstown, Northern Ireland, UK

Methods for Environmental Trace Analysis
John R. Dean, Northumbria University, Newcastle, UK

Liquid Chromatography−Mass Spectrometry: An Introduction
Robert E. Ardrey, University of Huddersfield, Huddersfield, UK

The Analysis of Controlled Substances
Michael D. Cole, Anglia Polytechnic University, Cambridge, UK

Infrared Spectroscopy: Fundamentals and Applications
Barbara H. Stuart, University of Technology, Sydney, Australia

Forthcoming Titles

Practical Inductively Coupled Plasma Spectroscopy
John R. Dean, Northumbria University, Newcastle, UK

Techniques of Modern Organic Mass Spectrometry
Robert E. Ardrey, University of Huddersfield, Huddersfield, UK

INFRARED SPECTROSCOPY: FUNDAMENTALS AND APPLICATIONS

Barbara H. Stuart

University of Technology, Sydney, Australia

John Wiley & Sons, Ltd

Other Wiley Editorial Offices

John Wiley & Sons Inc., 111 River Street, Hoboken, NJ 07030, USA

Jossey-Bass, 989 Market Street, San Francisco, CA 94103-1741, USA

Wiley-VCH Verlag GmbH, Boschstr. 12, D-69469 Weinheim, Germany

John Wiley & Sons Australia Ltd, 33 Park Road, Milton, Queensland 4064, Australia

John Wiley & Sons (Asia) Pte Ltd, 2 Clementi Loop #02-01, Jin Xing Distripark, Singapore 129809

John Wiley & Sons Canada Ltd, 22 Worcester Road, Etobicoke, Ontario, Canada M9W 1L1

Wiley also publishes its books in a variety of electronic formats. Some content that appears
in print may not be available in electronic books.

Library of Congress Cataloging-in-Publication Data

Stuart, Barbara (Barbara H.)
    Infrared spectroscopy : fundamentals and applications / Barbara H.
Stuart.
        p. cm.–(Analytical techniques in the sciences)
    Includes bibliographical references and index.
    ISBN 0-470-85427-8 (acid-free paper)–ISBN 0-470-85428-6 (pbk. :
acid-free paper)
    1. Infrared spectroscopy. I. Title. II. Series.
QD96.15S76 2004
543′.57–dc22

                                                  2003027960

British Library Cataloguing in Publication Data

A catalogue record for this book is available from the British Library

ISBN 0-470-85427-8 (Cloth)
ISBN 0-470-85428-6 (Paper)

Typeset in 10/12pt Times by Laserwords Private Limited, Chennai, India

This book is printed on acid-free paper responsibly manufactured from sustainable forestry
in which at least two trees are planted for each one used for paper production.

Contents

Series Preface

There has been a rapid expansion in the provision of further education in recent years, which has brought with it the need to provide more flexible methods of teaching in order to satisfy the requirements of an increasingly more diverse type of student. In this respect, the *open learning* approach has proved to be a valuable and effective teaching method, in particular for those students who for a variety of reasons cannot pursue full-time traditional courses. As a result, John Wiley & Sons, Ltd first published the Analytical Chemistry by Open Learning (ACOL) series of textbooks in the late 1980s. This series, which covers all of the major analytical techniques, rapidly established itself as a valuable teaching resource, providing a convenient and flexible means of studying for those people who, on account of their individual circumstances, were not able to take advantage of more conventional methods of education in this particular subject area.

Following upon the success of the ACOL series, which by its very name is predominately concerned with Analytical *Chemistry*, the *Analytical Techniques in the Sciences* (AnTS) series of open learning texts has been introduced with the aim of providing a broader coverage of the many areas of science in which analytical techniques and methods are now increasingly applied. With this in mind, the AnTS series of texts seeks to provide a range of books which will cover not only the actual techniques themselves, but *also* those scientific disciplines which have a necessary requirement for analytical characterization methods.

Analytical instrumentation continues to increase in sophistication, and as a consequence, the range of materials that can now be almost routinely analysed has increased accordingly. Books in this series which are concerned with the *techniques* themselves will reflect such advances in analytical instrumentation, while at the same time providing full and detailed discussions of the fundamental concepts and theories of the particular analytical method being considered. Such books will cover a variety of techniques, including general instrumental analysis, spectroscopy, chromatography, electrophoresis, tandem techniques, electroanalytical methods, X-ray analysis and other significant topics. In addition, books in the series will include the *application* of analytical techniques in areas such as environmental science, the life sciences, clinical analysis, food science, forensic analysis, pharmaceutical science, conservation and archaeology, polymer science and general solid-state materials science.

Written by experts in their own particular fields, the books are presented in an easy-to-read, user-friendly style, with each chapter including both learning objectives and summaries of the subject matter being covered. The progress of the reader can be assessed by the use of frequent self-assessment questions (SAQs) and discussion questions (DQs), along with their corresponding reinforcing or remedial responses, which appear regularly throughout the texts. The books are thus eminently suitable both for self-study applications and for forming the basis of industrial company in-house training schemes. Each text also contains a large amount of supplementary material, including bibliographies, lists of acronyms and abbreviations, and tables of SI Units and important physical constants, plus, where appropriate, glossaries and references to literature sources.

It is therefore hoped that this present series of textbooks will prove to be a useful and valuable source of teaching material, both for individual students and for teachers of science courses.

Dave Ando
Dartford, UK

Preface

Infrared spectroscopy is one of the most important and widely used analytical techniques available to scientists working in a whole range of fields. There are a number of texts on the subject available, ranging from instrumentation to specific applications. This present book aims to provide an introduction to those needing to use infrared spectroscopy for the first time, by explaining the fundamental aspects of the technique, how to obtain a spectrum and how to analyse infrared data obtained for a wide number of materials.

This text is not intended to be comprehensive, as infrared spectroscopy is extensively used. However, the information provided here may be used as a starting point for more detailed investigations. The book is laid out with introductory chapters covering the background theory of infrared spectroscopy, instrumentation and sampling techniques. Scientists may require qualitative and/or quantitative analysis of infrared data and therefore a chapter is devoted to the approaches commonly used to extract such information.

Infrared spectroscopy is a versatile experimental technique. It can be used to obtain important information about everything from delicate biological samples to tough minerals. In this book, the main areas that are studied using infrared spectroscopy are examined in a series of chapters, namely organic molecules, inorganic molecules, polymers, and biological, industrial and environmental applications. Each chapter provides examples of commonly encountered molecular structures in each field and how to approach the analysis of such structures. Suitable questions and problems are included in each chapter to assist in the analysis of the relevant infrared spectra.

I very much hope that those learning about and utilizing infrared spectroscopy will find this text a useful and valuable introduction to this major analytical technique.

Barbara Stuart
University of Technology, Sydney, Australia

Acronyms, Abbreviations and Symbols

ANN	artificial neural network
ATR	attenuated total reflectance
CLS	classical least-squares
D_2O	deuterium oxide
DAC	diamond anvil cell
DNA	deoxyribonucleic acid
DOP	dioctyl phthalate
DRIFT	diffuse reflectance infrared technique
DTGS	deuterium triglycine sulfate
EGA	evolved gas analysis
en	ethylenediamine
FFT	fast Fourier-transform
FPA	focal plane array
FTIR	Fourier-transform infrared (spectroscopy)
GC–IR	gas chromatography–infrared (spectroscopy)
GC–MS	gas chromatography–mass spectrometry
HDPE	high-density polyethylene
ILS	inverse least-squares
KRS-5	thallium-iodide
LC	liquid chromatography
LDA	linear discriminant analysis
LDPE	low-density polyethylene
MBP	myelin basic protein
MCT	mercury cadmium telluride
MIR	multiple internal reflectance

MMA	methyl methacrylate
NMR	nuclear magnetic resonance (spectroscopy)
PAS	photoacoustic spectroscopy
PCA	principal component analysis
PE	polyethylene
PEO	poly(ethylene oxide)
PET	poly(ethylene terephthalate)
PLS	partial least-squares
PMMA	poly(methyl methacrylate)
PP	polypropylene
PTFE	polytetrafluoroethylene
PU	polyurethane
PVC	poly(vinyl chloride)
PVIE	poly(vinyl isobutylether)
PVPh	poly(vinyl phenol)
RNA	ribonucleic acid
SFC	supercritical fluid chromatography
SNR	signal-to-noise ratio
TFE	trifluoroethanol
TGA	thermogravimetric analysis
TGA–IR	thermogravimetric analysis–infrared (spectroscopy)

A	absorbance
$A_{\parallel}$	absorbance parallel to chain axis
$A_{\perp}$	absorbance perpendicular to chain axis
B	magnetic vector (magnitude)
$B(\bar{\nu})$	spectral power density
c	speed of light; concentration
d_{p}	penetration depth
D	optical path difference
E	energy; electric vector (magnitude)
h	Planck constant
k	force constant; molar absorption coefficient
I	transmitted light
I_0	incident light
$I(\delta)$	intensity at detector
l	pathlength
L	cell pathlength
n	number of peak-to-peak fringes; refractive index; number of moles
P	pressure
R	reflectance; universal gas constant
T	transmittance; temperature

V	volume
δ	pathlength
ε	molar absorptivity
θ	angle of incident radiation
λ	wavelength
μ	reduced mass
ν	frequency
$\bar{\nu}$	wavenumber

About the Author

Barbara Stuart, B.Sc. (Sydney), M.Sc. (Sydney), Ph.D. (London), D.I.C., MRACI, MRSC, CChem

After graduating with a B.Sc. degree from the University of Sydney in Australia, Barbara Stuart then worked as a tutor at this university. She also carried out research in the field of biophysical chemistry in the Department of Physical Chemistry and graduated with an M.Sc. in 1990. The author then moved to the UK to carry out doctoral studies in polymer engineering within the Department of Chemical Engineering and Chemical Technology at Imperial College (University of London). After obtaining her Ph.D. in 1993, she took up a position as a Lecturer in Physical Chemistry at the University of Greenwich in South East London. Barbara returned to Australia in 1995, joining the staff at the University of Technology, Sydney, where she is currently a Senior Lecturer in the Department of Chemistry, Materials and Forensic Science. She is presently conducting research in the fields of polymer spectroscopy, materials conservation and forensic science. Barbara is the author of three other books published by John Wiley and Sons, Ltd, namely *Modern Infrared Spectroscopy* and *Biological Applications of Infrared Spectroscopy*, both in the ACOL series of open learning texts, and *Polymer Analysis* in this current AnTS series of texts.

Chapter 1
Introduction

Learning Objectives

- To understand the origin of electromagnetic radiation.
- To determine the frequency, wavelength, wavenumber and energy change associated with an infrared transition.
- To appreciate the factors governing the intensity of bands in an infrared spectrum.
- To predict the number of fundamental modes of vibration of a molecule.
- To understand the influences of force constants and reduced masses on the frequency of band vibrations.
- To appreciate the different possible modes of vibration.
- To recognize the factors that complicate the interpretation of infrared spectra.

Infrared spectroscopy is certainly one of the most important analytical techniques available to today's scientists. One of the great advantages of infrared spectroscopy is that virtually any sample in virtually any state may be studied. Liquids, solutions, pastes, powders, films, fibres, gases and surfaces can all be examined with a judicious choice of sampling technique. As a consequence of the improved instrumentation, a variety of new sensitive techniques have now been developed in order to examine formerly intractable samples.

Infrared spectrometers have been commercially available since the 1940s. At that time, the instruments relied on prisms to act as dispersive elements,

Infrared Spectroscopy: Fundamentals and Applications   B. Stuart
© 2004 John Wiley & Sons, Ltd   ISBNs: 0-470-85427-8 (HB); 0-470-85428-6 (PB)

but by the mid 1950s, diffraction gratings had been introduced into dispersive machines. The most significant advances in infrared spectroscopy, however, have come about as a result of the introduction of Fourier-transform spectrometers. This type of instrument employs an interferometer and exploits the well-established mathematical process of Fourier-transformation. Fourier-transform infrared (FTIR) spectroscopy has dramatically improved the quality of infrared spectra and minimized the time required to obtain data. In addition, with constant improvements to computers, infrared spectroscopy has made further great strides.

Infrared spectroscopy is a technique based on the vibrations of the atoms of a molecule. An infrared spectrum is commonly obtained by passing infrared radiation through a sample and determining what fraction of the incident radiation is absorbed at a particular energy. The energy at which any peak in an absorption spectrum appears corresponds to the frequency of a vibration of a part of a sample molecule. In this introductory chapter, the basic ideas and definitions associated with infrared spectroscopy will be described. The vibrations of molecules will be looked at here, as these are crucial to the interpretation of infrared spectra.

Once this chapter has been completed, some idea about the information to be gained from infrared spectroscopy should have been gained. The following chapter will aid in an understanding of how an infrared spectrometer produces a spectrum. After working through that chapter, it should be possible to record a spectrum and in order to do this a decision on an appropriate sampling technique needs to be made. The sampling procedure depends very much on the type of sample to be examined, for instance, whether it is a solid, liquid or gas. Chapter 2 also outlines the various sampling techniques that are commonly available. Once the spectrum has been recorded, the information it can provide needs to be extracted. Chapter 3, on spectrum interpretation, will assist in the understanding of the information to be gained from an infrared spectrum. As infrared spectroscopy is now used in such a wide variety of scientific fields, some of the many applications of the technique are examined in Chapters 4 to 8. These chapters should provide guidance as to how to approach a particular analytical problem in a specific field. The applications have been divided into separate chapters on organic and inorganic molecules, polymers, biological applications and industrial applications. This book is, of course, not meant to provide a comprehensive review of the use of infrared spectroscopy in each of these fields. However, an overview of the approaches taken in these areas is provided, along with appropriate references to the literature available in each of these disciplines.

1.1 Electromagnetic Radiation

The visible part of the electromagnetic spectrum is, by definition, radiation visible to the human eye. Other detection systems reveal radiation beyond the visible regions of the spectrum and these are classified as radiowave, microwave, infrared, ultraviolet, X-ray and γ-ray. These regions are illustrated in Figure 1.1, together with the processes involved in the interaction of the radiation of these regions with matter. The electromagnetic spectrum and the varied interactions between these radiations and many forms of matter can be considered in terms of either classical or quantum theories.

The nature of the various radiations shown in Figure 1.1 have been interpreted by Maxwell's classical theory of electro- and magneto-dynamics – hence, the term *electromagnetic radiation*. According to this theory, radiation is considered as two mutually perpendicular electric and magnetic fields, oscillating in single planes at right angles to each other. These fields are in phase and are being propagated as a sine wave, as shown in Figure 1.2. The magnitudes of the electric and magnetic vectors are represented by E and B, respectively.

A significant discovery made about electromagnetic radiation was that the velocity of propagation in a vacuum was constant for all regions of the spectrum. This is known as the velocity of light, c, and has the value $2.997\,925 \times 10^8$ m s^{-1}. If one complete wave travelling a fixed distance each cycle is visualized, it may be observed that the velocity of this wave is the product of the *wavelength*, λ (the distance between adjacent peaks), and the *frequency*, ν (the number of cycles

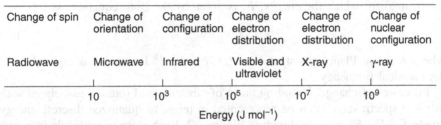

Change of spin	Change of orientation	Change of configuration	Change of electron distribution	Change of electron distribution	Change of nuclear configuration
Radiowave	Microwave	Infrared	Visible and ultraviolet	X-ray	γ-ray
	10	10³	10⁵	10⁷	10⁹

Energy (J mol⁻¹)

Figure 1.1 Regions of the electromagnetic spectrum. From Stuart, B., *Biological Applications of Infrared Spectroscopy*, ACOL Series, Wiley, Chichester, UK, 1997. © University of Greenwich, and reproduced by permission of the University of Greenwich.

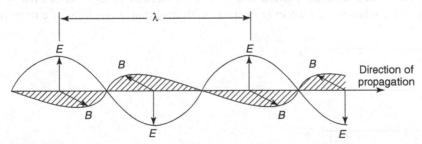

Figure 1.2 Representation of an electromagnetic wave. Reproduced from Brittain, E. F. H., George, W. O. and Wells, C. H. J., *Introduction to Molecular Spectroscopy*, Academic Press, London, Copyright (1975), with permission from Elsevier.

per second). Therefore:

$$c = \lambda v \tag{1.1}$$

The presentation of spectral regions may be in terms of wavelength as metres or sub-multiples of a metre. The following units are commonly encountered in spectroscopy:

$$1 \text{ Å} = 10^{-10} \text{ m} \qquad 1 \text{ nm} = 10^{-9} \text{ m} \qquad 1 \text{ μm} = 10^{-6} \text{ m}$$

Another unit which is widely used in infrared spectroscopy is the *wavenumber*, $\bar{v}$, in cm^{-1}. This is the number of waves in a length of one centimetre and is given by the following relationship:

$$\bar{v} = 1/\lambda = v/c \tag{1.2}$$

This unit has the advantage of being linear with energy.

During the 19th Century, a number of experimental observations were made which were not consistent with the classical view that matter could interact with energy in a continuous form. Work by Einstein, Planck and Bohr indicated that in many ways electromagnetic radiation could be regarded as a stream of particles (or quanta) for which the energy, E, is given by the Bohr equation, as follows:

$$E = hv \tag{1.3}$$

where h is the Planck constant ($h = 6.626 \times 10^{-34}$ J s) and v is equivalent to the classical frequency.

Processes of change, including those of vibration and rotation associated with infrared spectroscopy, can be represented in terms of quantized discrete energy levels E_0, E_1, E_2, etc., as shown in Figure 1.3. Each atom or molecule in a system must exist in one or other of these levels. In a large assembly of molecules, there will be a distribution of all atoms or molecules among these various energy levels. The latter are a function of an integer (the *quantum number*) and a parameter associated with the particular atomic or molecular process associated with that state. Whenever a molecule interacts with radiation, a quantum of energy (or

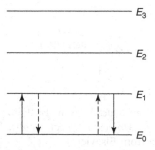

Figure 1.3 Illustration of quantized discrete energy levels.

photon) is either emitted or absorbed. In each case, the energy of the quantum of radiation must exactly fit the energy gap $E_1 - E_0$ or $E_2 - E_1$, etc. The energy of the quantum is related to the frequency by the following:

$$\Delta E = h\nu \qquad (1.4)$$

Hence, the frequency of emission or absorption of radiation for a transition between the energy states E_0 and E_1 is given by:

$$\nu = (E_1 - E_0)/h \qquad (1.5)$$

Associated with the uptake of energy of quantized absorption is some deactivation mechanism whereby the atom or molecule returns to its original state. Associated with the loss of energy by emission of a quantum of energy or photon is some prior excitation mechanism. Both of these associated mechanisms are represented by the dotted lines in Figure 1.3.

SAQ 1.1

Caffeine molecules absorb infrared radiation at 1656 cm^{-1}. Calculate the following:

(i) wavelength of this radiation;

(ii) frequency of this radiation;

(iii) energy change associated with this absorption.

1.2 Infrared Absorptions

For a molecule to show infrared absorptions it must possess a specific feature, i.e. an electric dipole moment of the molecule must change during the vibration. This is the *selection rule* for infrared spectroscopy. Figure 1.4 illustrates an example of an 'infrared-active' molecule, a *heteronuclear* diatomic molecule. The dipole moment of such a molecule changes as the bond expands and contracts. By comparison, an example of an 'infrared-inactive' molecule is a *homonuclear* diatomic molecule because its dipole moment remains zero no matter how long the bond.

An understanding of molecular symmetry and group theory is important when initially assigning infrared bands. A detailed description of such theory is beyond the scope of this book, but symmetry and group theory are discussed in detail in other texts [1, 2]. Fortunately, it is not necessary to work from first principles each time a new infrared spectrum is obtained.

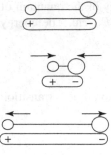

Figure 1.4 Change in the dipole moment of a heteronuclear diatomic molecule.

Infrared absorptions are not infinitely narrow and there are several factors that contribute to the broadening. For gases, the Doppler effect, in which radiation is shifted in frequency when the radiation source is moving towards or away from the observer, is a factor. There is also the broadening of bands due to the collisions between molecules. Another source of line broadening is the finite lifetime of the states involved in the transition. From quantum mechanics, when the Schrödinger equation is solved for a system which is changing with time, the energy states of the system do not have precisely defined energies and this leads to lifetime broadening. There is a relationship between the lifetime of an excited state and the bandwidth of the absorption band associated with the transition to the excited state, and this is a consequence of the *Heisenberg Uncertainty Principle*. This relationship demonstrates that the shorter the lifetime of a state, then the less well defined is its energy.

1.3 Normal Modes of Vibration

The interactions of infrared radiation with matter may be understood in terms of changes in molecular dipoles associated with vibrations and rotations. In order to begin with a basic model, a molecule can be looked upon as a system of masses joined by bonds with spring-like properties. Taking first the simple case of diatomic molecules, such molecules have three degrees of translational freedom and two degrees of rotational freedom. The atoms in the molecules can also move relative to one other, that is, bond lengths can vary or one atom can move out of its present plane. This is a description of stretching and bending movements that are collectively referred to as *vibrations*. For a diatomic molecule, only one vibration that corresponds to the stretching and compression of the bond is possible. This accounts for one degree of vibrational freedom.

Polyatomic molecules containing many (N) atoms will have $3N$ degrees of freedom. Looking first at the case of molecules containing three atoms, two groups of triatomic molecules may be distinguished, i.e. linear and non-linear. Two simple examples of linear and non-linear triatomics are represented by CO_2

O
/ \   O = C = O
H   H

Non-linear   Linear   **Figure 1.5** Carbon dioxide and water molecules.

Table 1.1 Degrees of freedom for polyatomic molecules. From Stuart, B., *Modern Infrared Spectroscopy*, ACOL Series, Wiley, Chichester, UK, 1996. © University of Greenwich, and reproduced by permission of the University of Greenwich

Type of degrees of freedom	Linear	Non-linear
Translational	3	3
Rotational	2	3
Vibrational	$3N - 5$	$3N - 6$
Total	$3N$	$3N$

and H_2O, respectively (illustrated in Figure 1.5). Both CO_2 and H_2O have three degrees of translational freedom. Water has three degrees of rotational freedom, but the linear molecule carbon dioxide has only two since no detectable energy is involved in rotation around the O=C=O axis. Subtracting these from $3N$, there are $3N-5$ degrees of freedom for CO_2 (or any linear molecule) and $3N-6$ for water (or any non-linear molecule). N in both examples is three, and so CO_2 has four vibrational modes and water has three. The degrees of freedom for polyatomic molecules are summarized in Table 1.1.

SAQ 1.2

How many vibrational degrees of freedom does a chloroform ($CHCl_3$) molecule possess?

Whereas a diatomic molecule has only one mode of vibration which corresponds to a stretching motion, a non-linear B–A–B type triatomic molecule has three modes, two of which correspond to stretching motions, with the remainder corresponding to a bending motion. A linear type triatomic has four modes, two of which have the same frequency, and are said to be *degenerate*.

Two other concepts are also used to explain the frequency of vibrational modes. These are the stiffness of the bond and the masses of the atoms at each end of the bond. The stiffness of the bond can be characterized by a proportionality constant termed the *force constant*, k (derived from Hooke's law). The *reduced mass*, μ, provides a useful way of simplifying our calculations by combining the individual atomic masses, and may be expressed as follows:

$$(1/\mu) = (1/m_1) + (1/m_2) \tag{1.6}$$

where m_1 and m_2 are the masses of the atoms at the ends of the bond. A practical alternative way of expressing the reduced mass is:

$$\mu = m_1m_2/(m_1 + m_2) \tag{1.7}$$

The equation relating the force constant, the reduced mass and the frequency of absorption is:

$$\nu = (1/2\pi)\sqrt{(k/\mu)} \tag{1.8}$$

This equation may be modified so that direct use of the wavenumber values for bond vibrational frequencies can be made, namely:

$$\bar{\nu} = (1/2\pi c)\sqrt{(k/\mu)} \tag{1.9}$$

where c is the speed of light.

A molecule can only absorb radiation when the incoming infrared radiation is of the same frequency as one of the fundamental modes of vibration of the molecule. This means that the vibrational motion of a small part of the molecule is increased while the rest of the molecule is left unaffected.

SAQ 1.3

Given that the C–H stretching vibration for chloroform occurs at 3000 cm^{-1}, calculate the C–D stretching frequency for deuterochloroform. The relevant atomic masses are ^{1}H $= 1.674 \times 10^{-27}$ kg, ^{2}H $= 3.345 \times 10^{-27}$ kg and ^{12}C $= 1.993 \times 10^{-27}$ kg.

Vibrations can involve either a change in bond length (*stretching*) or bond angle (*bending*) (Figure 1.6). Some bonds can stretch in-phase (*symmetrical* stretching) or out-of-phase (*asymmetric* stretching), as shown in Figure 1.7. If a molecule has different terminal atoms such as HCN, ClCN or ONCl, then the two stretching modes are no longer symmetric and asymmetric vibrations of similar bonds, but will have varying proportions of the stretching motion of each group. In other words, the amount of *coupling* will vary.

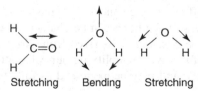

Figure 1.6 Stretching and bending vibrations.

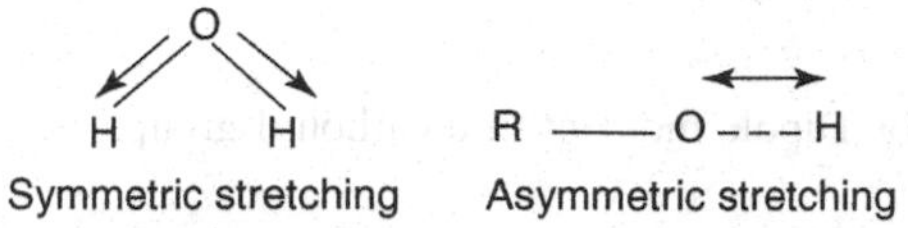

O
/   \
H   H   R ———— O --- H
Symmetric stretching   Asymmetric stretching

Figure 1.7 Symmetric and asymmetric stretching vibrations.

Bending vibrations also contribute to infrared spectra and these are summarized in Figure 1.8. It is best to consider the molecule being cut by a plane
through the hydrogen atoms and the carbon atom. The hydrogens can move in
the same direction or in opposite directions in this plane, here the plane of the
page. For more complex molecules, the analysis becomes simpler since hydrogen
atoms may be considered in isolation because they are usually attached to more
massive, and therefore, more rigid parts of the molecule. This results in *in-plane*
and *out-of-plane* bending vibrations, as illustrated in Figure 1.9.

As already mentioned, for a vibration to give rise to the absorption of infrared
radiation, it must cause a change in the dipole moment of the molecule. The
larger this change, then the more intense will be the absorption band. Because
of the difference in electronegativity between carbon and oxygen, the carbonyl
group is permanently polarized, as shown in Figure 1.10. Stretching this bond
will increase the dipole moment and, hence, $C{=}O$ stretching is an intense absorption. In CO_2, two different stretching vibrations are possible: (a) symmetric and
(b) asymmetric (Figure 1.11). In practice, this 'black and white' situation does
not prevail. The change in dipole may be very small and, hence, lead to a very
weak absorption.

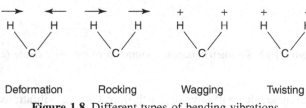

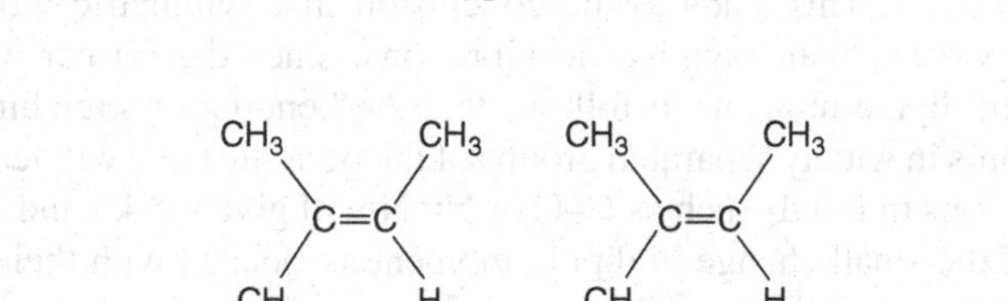

Deformation   Rocking   Wagging   Twisting

Figure 1.8 Different types of bending vibrations.

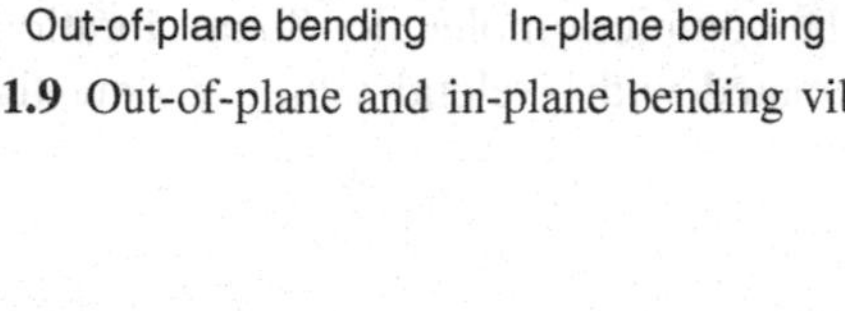

CH₃   CH₃   CH₃   CH₃
\   /   \   /
C=C   C=C
/   \   /   \
CH₃   H   CH₃   H
+

Out-of-plane bending   In-plane bending

Figure 1.9 Out-of-plane and in-plane bending vibrations.

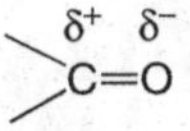

$\overset{\delta^+}{\underset{}{\diagdown}}\overset{\delta^-}{C{=}O}$ **Figure 1.10** Dipole moment in a carbonyl group.

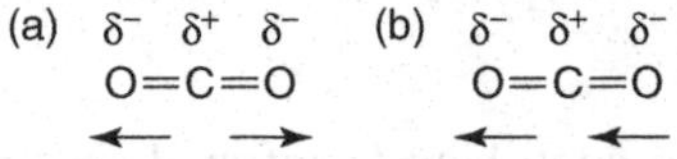

(a) $\delta^-\ \ \delta^+\ \ \delta^-$ (b) $\delta^-\ \ \delta^+\ \ \delta^-$
O=C=O   O=C=O
$\leftarrow\quad\rightarrow$ $\leftarrow\quad\leftarrow$

Figure 1.11 Stretching vibrations of carbon dioxide.

DQ 1.1

Which one of the vibrations shown in Figure 1.11 is 'infrared-inactive'?

Answer

A dipole moment is a vector sum. CO_2 in the ground state, therefore, has no dipole moment. If the two C=O bonds are stretched symmetrically, there is still no net dipole and so there is no infrared activity. However, in the asymmetric stretch, the two C=O bonds are of different length and, hence, the molecule has a dipole. Therefore, the vibration shown in Figure 1.11(b) is 'infrared-active'.

SAQ 1.4

Consider the symmetrical bending vibration of CO_2, as shown in Figure 1.12. Will this vibration be 'active' in the infrared?

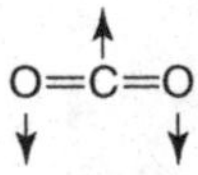

Figure 1.12 Symmetric bending vibration of carbon dioxide (cf. SAQ 1.4).

Symmetrical molecules will have fewer 'infrared-active' vibrations than asymmetrical molecules. This leads to the conclusion that symmetric vibrations will generally be weaker than asymmetric vibrations, since the former will not lead to a change in dipole moment. It follows that the bending or stretching of bonds involving atoms in widely separated groups of the periodic table will lead to intense bands. Vibrations of bonds such as C–C or N=N will give weak bands. This again is because of the small change in dipole moment associated with their vibrations.

There will be many different vibrations for even fairly simple molecules. The complexity of an infrared spectrum arises from the coupling of vibrations over a large part of or over the complete molecule. Such vibrations are called *skeletal* vibrations. Bands associated with skeletal vibrations are likely to conform to a pattern or *fingerprint* of the molecule as a whole, rather than a specific group within the molecule.

1.4 Complicating Factors

There are a number of factors that may complicate the interpretation of infrared spectra. These factors should be considered when studying spectra as they can result in important changes to the spectra and may result in the misinterpretation of bands.

1.4.1 Overtone and Combination Bands

The sound we hear is a mixture of harmonics, that is, a fundamental frequency mixed with multiples of that frequency. *Overtone bands* in an infrared spectrum are analogous and are multiples of the fundamental absorption frequency. The energy levels for overtones of infrared modes are illustrated in Figure 1.13. The energy required for the first overtone is twice the fundamental, assuming evenly spaced energy levels. Since the energy is proportional to the frequency absorbed and this is proportional to the wavenumber, the first overtone will appear in the spectrum at twice the wavenumber of the fundamental.

Combination bands arise when two fundamental bands absorbing at $\bar{v}_1$ and $\bar{v}_2$ absorb energy simultaneously. The resulting band will appear at $(\bar{v}_1 + \bar{v}_2)$ wavenumbers.

SAQ 1.5

A molecule has strong fundamental bands at the following wavenumbers:

C–H bending at 730 cm^{-1}

C–C stretching at 1400 cm^{-1}

C–H stretching at 2950 cm^{-1}

Determine the wavenumbers of the possible combination bands and the first overtones.

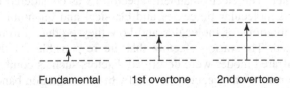

Fundamental   1st overtone   2nd overtone

Figure 1.13 Energy levels for fundamental and overtone infrared bands.

1.4.2 Fermi Resonance

The Fermi resonance effect usually leads to two bands appearing close together when only one is expected. When an overtone or a combination band has the same frequency as, or a similar frequency to, a fundamental, two bands appear, split either side of the expected value and are of about equal intensity. The effect is greatest when the frequencies match, but it is also present when there is a mismatch of a few tens of wavenumbers. The two bands are referred to as a *Fermi doublet*.

1.4.3 Coupling

Vibrations in the skeletons of molecules become coupled, as mentioned in Section 1.4. Such vibrations are not restricted to one or two bonds, but may involve a large part of the carbon backbone and oxygen or nitrogen atoms if present. The energy levels mix, hence resulting in the same number of vibrational modes, but at different frequencies, and bands can no longer be assigned to one bond. This is very common and occurs when adjacent bonds have similar frequencies. Coupling commonly occurs between C–C stretching, C–O stretching, C–N stretching, C–H rocking and C–H wagging motions. A further requirement is that to be strongly coupled, the motions must be in the same part of the molecule.

1.4.4 Vibration–Rotation Bands

When the infrared spectra of gaseous heteronuclear molecules are analysed at high resolution, a series of closely spaced components are observed. This type of structure is due to the excitation of rotational motion during a vibrational transition and is referred to as an vibration–rotation spectrum [1]. The absorptions fall into groups called branches and are labelled P, Q and R according to the change in the rotational quantum number associated with the transition. The separation of the lines appearing in a vibration–rotation spectrum may be exploited to determine the bond length of the molecule being examined.

Summary

The ideas fundamental to an understanding of infrared spectroscopy were introduced in this chapter. The electromagnetic spectrum was considered in terms of various atomic and molecular processes and classical and quantum ideas were introduced. The vibrations of molecules and how they produce infrared spectra were then examined. The various factors that are responsible for the position and intensity of infrared modes were described. Factors such as combination and overtone bands, Fermi resonance, coupling and vibration–rotation bands can lead to changes in infrared spectra. An appreciation of these issues is important when examining spectra and these factors were outlined in this chapter. For further reference, there is a range of books and book chapters available which provide an overview of the theory behind infrared spectroscopy [3–7].

References

1. Atkins, P. and de Paula, J., *Physical Chemistry*, 7th Edn, Oxford University Press, Oxford, UK, 2002.
2. Vincent, A., *Molecular Symmetry and Group Theory*, 2nd Edn, Wiley, Chichester, UK, 2001.
3. Günzler, H. and Gremlich, H.-U., *IR Spectroscopy: An Introduction*, Wiley-VCH, Weinheim, Germany, 2002.
4. Hollas, J. M., *Basic Atomic and Molecular Spectroscopy*, Wiley, Chichester, UK, 2002.
5. Steele, D., 'Infrared Spectroscopy: Theory', in *Handbook of Vibrational Spectroscopy*, Vol. 1, Chalmers, J. M. and Griffiths, P. R. (Eds), Wiley, Chichester, UK, 2002, pp. 44–70.
6. Barrow, G. M., *Introduction to Molecular Spectroscopy*, McGraw-Hill, New York, 1962.
7. Hollas, J. M., *Modern Spectroscopy*, 3rd Edn, Wiley, Chichester, UK, 1996.

Chapter 2
Experimental Methods

Learning Objectives

- To understand how an infrared spectrum is obtained from a Fourier-transform instrument.
- To recognize the different methods of sample preparation and sample handling techniques which are used for preparing samples in infrared spectroscopy.
- To recognize poor quality spectra and diagnose their causes.
- To understand the origins of reflectance techniques.
- To understand the origins of infrared microsampling techniques.
- To understand that spectral information may be obtained from combination infrared spectroscopy techniques.
- To select appropriate sample preparation methods for different types of samples.

2.1 Introduction

Traditionally, dispersive instruments, available since the 1940s, were used to obtain infrared spectra. In recent decades, a very different method of obtaining an infrared spectrum has superceded the dispersive instrument. Fourier-transform infrared spectrometers are now predominantly used and have improved the acquisition of infrared spectra dramatically. In this present chapter, the instrumentation required to obtain an infrared spectrum will be described.

Infrared spectroscopy is a versatile experimental technique and it is relatively easy to obtain spectra from samples in solution or in the liquid, solid or gaseous

Infrared Spectroscopy: Fundamentals and Applications   B. Stuart
© 2004 John Wiley & Sons, Ltd   ISBNs: 0-470-85427-8 (HB); 0-470-85428-6 (PB)

states. In this chapter, how samples can be introduced into the instrument, the equipment required to obtain spectra and the pre-treatment of samples are examined. First, the various ways of investigating samples using the traditional transmission methods of infrared spectroscopy will be discussed. Reflectance methods, such as the attenuated total reflectance, diffuse reflectance and specular reflectance approaches, as well as photoacoustic spectroscopy, are also explained. Infrared microspectroscopy has emerged in recent years as an effective tool for examining small and/or complex samples; the techniques used are described in this chapter. Infrared spectroscopy has also been combined with other well-established analytical techniques such as chromatography and thermal analysis. Such combination techniques are introduced here.

2.2 Dispersive Infrared Spectrometers

The first dispersive infrared instruments employed prisms made of materials such as sodium chloride. The popularity of prism instruments fell away in the 1960s when the improved technology of grating construction enabled cheap, good-quality gratings to be manufactured.

The dispersive element in dispersive instruments is contained within a monochromator. Figure 2.1 shows the optical path of an infrared spectrometer which uses a grating monochromator. Dispersion occurs when energy falling on the entrance slit is collimated onto the dispersive element and the dispersed radiation is then reflected back to the exit slit, beyond which lies the detector. The dispersed spectrum is scanned across the exit slit by rotating a suitable component within the monochromator. The widths of the entrance and exit slits may be varied and programmed to compensate for any variation of the source energy with wavenumber. In the absence of a sample, the detector then receives radiation of approximately constant energy as the spectrum is scanned.

Atmospheric absorption by CO_2 and H_2O in the instrument beam has to be considered in the design of infrared instruments. Figure 2.2 shows the spectrum of such atmospheric absorptions. These contributions can be taken into account by using a double-beam arrangement in which radiation from a source is divided into two beams. These beams pass through a sample and a reference path of the sample compartment, respectively. The information from these beams is rationed to obtain the required sample spectrum.

A detector must have adequate sensitivity to the radiation arriving from the sample and monochromator over the entire spectral region required. In addition, the source must be sufficiently intense over the wavenumber range and transmittance range. Sources of infrared emission have included the Globar, which is constructed of silicon carbide. There is also the Nernst filament, which is a mixture of the oxides of zirconium, yttrium and erbium. A Nernst filament only conducts electricity at elevated temperatures. Most detectors have consisted of thermocouples of varying characteristics.

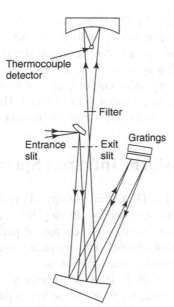

Figure 2.1 Schematic of the optical path of a double-beam infrared spectrometer with a grating monochromator. Reproduced from Brittain, E. F. H., George, W. O. and Wells, C. H. J., *Introduction to Molecular Spectroscopy*, Academic Press, London, Copyright (1975), with permission from Elsevier.

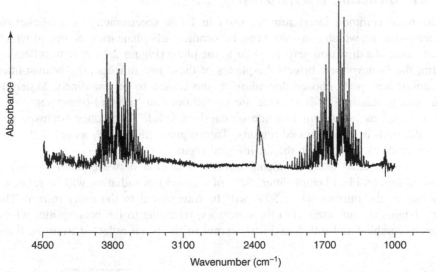

Figure 2.2 Infrared spectrum of atmospheric contributions (e.g. CO_2 and H_2O). From Stuart, B., *Modern Infrared Spectroscopy*, ACOL Series, Wiley, Chichester, UK, 1996. © University of Greenwich, and reproduced by permission of the University of Greenwich.

The essential problem of the dispersive spectrometer lies with its monochromator. This contains narrow slits at the entrance and exit which limit the wavenumber range of the radiation reaching the detector to one resolution width. Samples for which a very quick measurement is needed, for example, in the eluant from a chromatography column, cannot be studied with instruments of low sensitivity because they cannot scan at speed. However, these limitations may be overcome through the use of a Fourier-transform infrared spectrometer.

2.3 Fourier-Transform Infrared Spectrometers

Fourier-transform infrared (FTIR) spectroscopy [1] is based on the idea of the interference of radiation between two beams to yield an *interferogram*. The latter is a signal produced as a function of the change of pathlength between the two beams. The two domains of distance and frequency are interconvertible by the mathematical method of *Fourier-transformation*.

The basic components of an FTIR spectrometer are shown schematically in Figure 2.3. The radiation emerging from the source is passed through an interferometer to the sample before reaching a detector. Upon amplification of the signal, in which high-frequency contributions have been eliminated by a filter, the data are converted to digital form by an analog-to-digital converter and transferred to the computer for Fourier-transformation.

2.3.1 Michelson Interferometers

The most common interferometer used in FTIR spectrometry is a Michelson interferometer, which consists of two perpendicularly plane mirrors, one of which can travel in a direction perpendicular to the plane (Figure 2.4). A semi-reflecting film, the *beamsplitter*, bisects the planes of these two mirrors. The beamsplitter material has to be chosen according to the region to be examined. Materials such as germanium or iron oxide are coated onto an 'infrared-transparent' substrate such as potassium bromide or caesium iodide to produce beamsplitters for the mid- or near-infrared regions. Thin organic films, such as poly(ethylene terephthalate), are used in the far-infrared region.

If a collimated beam of monochromatic radiation of wavelength λ (cm) is passed into an ideal beamsplitter, 50% of the incident radiation will be reflected to one of the mirrors while 50% will be transmitted to the other mirror. The two beams are reflected from these mirrors, returning to the beamsplitter where they recombine and interfere. Fifty percent of the beam reflected from the fixed

Figure 2.3 Basic components of an FTIR spectrometer.

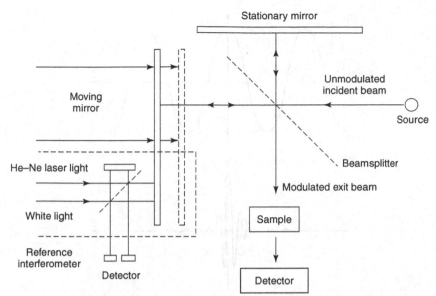

Figure 2.4 Schematic of a Michelson interferometer. From Stuart, B., *Modern Infrared Spectroscopy*, ACOL Series, Wiley, Chichester, UK, 1996. © University of Greenwich, and reproduced by permission of the University of Greenwich.

mirror is transmitted through the beamsplitter while 50% is reflected back in the direction of the source. The beam which emerges from the interferometer at 90° to the input beam is called the transmitted beam and this is the beam detected in FTIR spectrometry.

The moving mirror produces an optical path difference between the two arms of the interferometer. For path differences of $(n + 1/2)\lambda$, the two beams interfere destructively in the case of the transmitted beam and constructively in the case of the reflected beam. The resultant interference pattern is shown in Figure 2.5 for (a) a source of monochromatic radiation and (b) a source of polychromatic radiation (b). The former is a simple cosine function, but the latter is of a more complicated form because it contains all of the spectral information of the radiation falling on the detector.

2.3.2 Sources and Detectors

FTIR spectrometers use a Globar or Nernst source for the mid-infrared region. If the far-infrared region is to be examined, then a high-pressure mercury lamp can be used. For the near-infrared, tungsten–halogen lamps are used as sources.

There are two commonly used detectors employed for the mid-infrared region. The normal detector for routine use is a pyroelectric device incorporating deuterium tryglycine sulfate (DTGS) in a temperature-resistant alkali halide window.

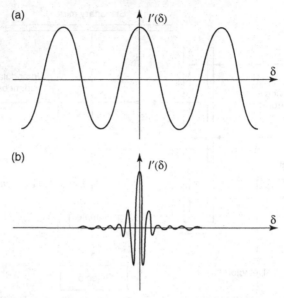

Figure 2.5 Interferograms obtained for (a) monochromatic radiation and (b) polychromatic radiation. Reproduced with permission from Barnes, A. J. and Orville-Thomas, W. J. (Eds), *Vibrational Spectroscopy – Modern Trends*, Elsevier, Amsterdam, Figure 2, p. 55 (1977).

For more sensitive work, mercury cadmium telluride (MCT) can be used, but this has to be cooled to liquid nitrogen temperatures. In the far-infrared region, germanium or indium–antimony detectors are employed, operating at liquid helium temperatures. For the near-infrared region, the detectors used are generally lead sulfide photoconductors.

2.3.3 Fourier-Transformation

The essential equations for a Fourier-transformation relating the intensity falling on the detector, $I(\delta)$, to the spectral power density at a particular wavenumber, $\bar{\nu}$, given by $B(\bar{\nu})$, are as follows:

$$I(\delta) = \int_{0}^{+\infty} B(\bar{\nu})\cos{(2\pi\bar{\nu}\delta)}d\bar{\nu} \qquad (2.1)$$

which is one half of a cosine Fourier-transform pair, with the other being:

$$B(\bar{\nu}) = \int_{-\infty}^{+\infty} I(\delta)\cos{(2\pi\bar{\nu}\delta)}d\delta \qquad (2.2)$$

These two equations are interconvertible and are known as a Fourier-transform pair. The first shows the variation in power density as a function of the difference in pathlength, which is an interference pattern. The second shows the variation in intensity as a function of wavenumber. Each can be converted into the other by the mathematical method of *Fourier-transformation*.

The essential experiment to obtain an FTIR spectrum is to produce an interferogram with and without a sample in the beam and transforming the interferograms into spectra of (a) the source with sample absorptions and (b) the source without sample absorptions. The ratio of the former and the latter corresponds to a double-beam dispersive spectrum.

The major advance toward routine use in the mid-infrared region came with a new mathematical method (or algorithm) devised for *fast Fourier-transformation* (FFT). This was combined with advances in computers which enabled these calculations to be carried out rapidly.

2.3.4 Moving Mirrors

The moving mirror is a crucial component of the interferometer. It has to be accurately aligned and must be capable of scanning two distances so that the path difference corresponds to a known value. A number of factors associated with the moving mirror need to be considered when evaluating an infrared spectrum.

The interferogram is an analogue signal at the detector that has to be digitized in order that the Fourier-transformation into a conventional spectrum can be carried out. There are two particular sources of error in transforming the digitized information on the interferogram into a spectrum. First, the transformation carried out in practice involves an integration stage over a finite displacement rather than over an infinite displacement. The mathematical process of Fourier-transformation assumes infinite boundaries. The consequence of this necessary approximation is that the apparent lineshape of a spectral line may be as shown in Figure 2.6, where the main band area has a series of negative and positive side lobes (or pods) with diminishing amplitudes.

The process of *apodization* is the removal of the side lobes (or pods) by multiplying the interferogram by a suitable function before the Fourier-transformation is carried out. A suitable function must cause the intensity of the interferogram to fall smoothly to zero at its ends. Most FTIR spectrometers offer a choice of apodization options and a good general purpose apodization function is the cosine function, as follows:

$$F(D) = [1 + \cos(\pi D)]/2 \qquad (2.3)$$

where D is the optical path difference. This cosine function provides a good compromise between reduction in oscillations and deterioration in spectral resolution. When accurate band shapes are required, more sophisticated mathematical functions may be needed.

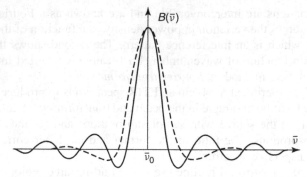

Figure 2.6 Instrument lineshape without apodization. Reproduced with permission from Barnes, A. J. and Orville-Thomas, W. J. (Eds), *Vibrational Spectroscopy – Modern Trends*, Elsevier, Amsterdam, Figure 3, p. 55 (1977).

Another source of error arises if the sample intervals are not exactly the same on each side of the maxima corresponding to zero path differences. Phase correction is required and this correction procedure ensures that the sample intervals are the same on each side of the first interval and should correspond to a path difference of zero.

The resolution for an FTIR instrument is limited by the maximum path difference between the two beams. The limiting resolution in wavenumbers (cm^{-1}) is the reciprocal of the pathlength difference (cm). For example, a pathlength difference of 10 cm is required to achieve a limiting resolution of 0.1 cm^{-1}. This simple calculation appears to show that it is easy to achieve high resolution. Unfortunately, this is not the case since the precision of the optics and mirror movement mechanism become more difficult to achieve at longer displacements of pathlengths.

SAQ 2.1

An FTIR spectrometer is used to record a single-beam spectrum from a single scan with a difference in pathlength (δ) of 100 mm.

(a) What is the limiting resolution in units of cm^{-1}?

(b) How could a limiting resolution of 0.02 cm^{-1} be achieved?

2.3.5 Signal-Averaging

The main advantage of rapid-scanning instruments is the ability to increase the signal-to-noise ratio (SNR) by signal-averaging, leading to an increase of signal-to-noise proportional to the square root of the time, as follows:

$$\text{SNR} \; \alpha \; n^{1/2} \tag{2.4}$$

There are diminishing returns for signal-averaging in that it takes an increasingly longer time to achieve greater and greater improvement. The accumulation of a large number of repeat scans makes greater demands on the instrument if it is to exactly reproduce the conditions. It is normal to incorporate a laser monochromatic source in the beam of the continuous source. The laser beam produces standard fringes which can 'line-up' successive scans accurately and can determine and control the displacement of the moving mirror at all times.

2.3.6 Advantages

FTIR instruments have several significant advantages over older dispersive instruments. Two of these are the Fellgett (or multiplex) advantage and the Jacquinot (or throughput) advantage. The *Fellgett advantage* is due to an improvement in the SNR per unit time, proportional to the square root of the number of resolution elements being monitored. This results from the large number of resolution elements being monitored simultaneously. In addition, because FTIR spectrometry does not require the use of a slit or other restricting device, the total source output can be passed through the sample continuously. This results in a substantial gain in energy at the detector, hence translating to higher signals and improved SNRs. This is known as *Jacquinot's advantage*.

Another strength of FTIR spectrometry is its *speed advantage*. The mirror has the ability to move short distances quite rapidly, and this, together with the SNR improvements due to the Fellgett and Jacquinot advantages, make it possible to obtain spectra on a millisecond timescale. In interferometry, the factor which determines the precision of the position of an infrared band is the precision with which the scanning mirror position is known. By using a helium—neon laser as a reference, the mirror position is known with high precision.

2.3.7 Computers

The computer forms a crucial component of modern infrared instruments and performs a number of functions. The computer controls the instrument, for example, it sets scan speeds and scanning limits, and starts and stops scanning. It reads spectra into the computer memory from the instrument as the spectrum is scanned; this means that the spectrum is digitized. Spectra may be manipulated using the computer, for example, by adding and subtracting spectra or expanding areas of the spectrum of interest. The computer is also used to scan the spectra continuously and average or add the result in the computer memory. Complex analyses may be automatically carried out by following a set of pre-programmed commands (described later in Chapter 3). The computer is also used to plot the spectra.

2.3.8 Spectra

Early infrared instruments recorded percentage transmittance over a linear wavelength range. It is now unusual to use wavelength for routine samples and the wavenumber scale is commonly used. The output from the instrument is referred to as a *spectrum*. Most commercial instruments present a spectrum with the wavenumber decreasing from left to right.

The infrared spectrum can be divided into three main regions: the *far-infrared* (<400 cm^{-1}), the *mid-infrared* ($4000-400$ cm^{-1}) and the *near-infrared* ($13\,000-4000$ cm^{-1}). These regions will be described later in more detail in Chapter 3. Many infrared applications employ the mid-infrared region, but the near- and far-infrared regions also provide important information about certain materials. Generally, there are less infrared bands in the $4000-1800$ cm^{-1} region with many bands between 1800 and 400 cm^{-1}. Sometimes, the scale is changed so that the region between 4000 and 1800 cm^{-1} is contracted and the region between 1800 and 400 cm^{-1} is expanded to emphasize features of interest.

The ordinate scale may be presented in % transmittance with 100% at the top of the spectrum. It is commonplace to have a choice of absorbance or transmittance as a measure of band intensity. The relationship between these two quantities will be described in Chapter 3. Figures 2.7 and 2.8 show the infrared spectra of lactic acid and illustrate the difference in appearance between absorbance and

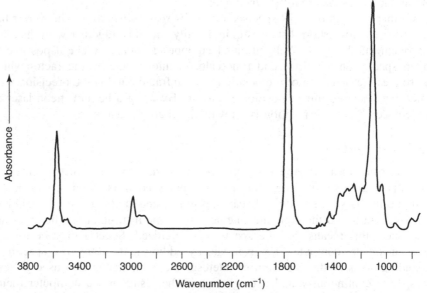

Figure 2.7 Absorbance spectrum of lactic acid. From Stuart, B., *Biological Applications of Infrared Spectroscopy*, ACOL Series, Wiley, Chichester, UK, 1997. © University of Greenwich, and reproduced by permission of the University of Greenwich.

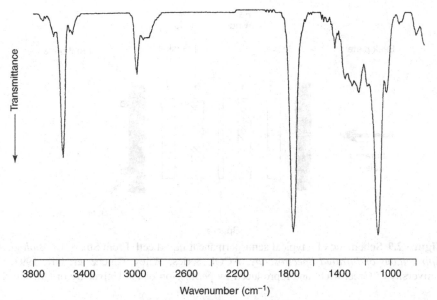

Figure 2.8 Transmittance spectrum of lactic acid. From Stuart, B., *Biological Applications of Infrared Spectroscopy*, ACOL Series, Wiley, Chichester, UK, 1997. © University of Greenwich, and reproduced by permission of the University of Greenwich.

transmittance spectra. It almost comes down to personal preference which of the two modes to use, but the transmittance is traditionally used for spectral interpretation, while absorbance is used for quantitative work.

2.4 Transmission Methods

Transmission spectroscopy is the oldest and most straightforward infrared method. This technique is based upon the absorption of infrared radiation at specific wavelengths as it passes through a sample. It is possible to analyse samples in the liquid, solid or gaseous forms when using this approach.

2.4.1 Liquids and Solutions

There are several different types of transmission solution cells available. Fixed-pathlength sealed cells are useful for volatile liquids, but cannot be taken apart for cleaning. Semi-permanent cells are demountable so that the windows can be cleaned. A semi-permanent cell is illustrated in Figure 2.9. The spacer is usually made of polytetrafluoroethylene (PTFE, known as 'Teflon') and is available in a variety of thicknesses, hence allowing one cell to be used for various pathlengths. Variable pathlength cells incorporate a mechanism for continuously adjusting the pathlength, while a vernier scale allows accurate adjustment. All of these cell

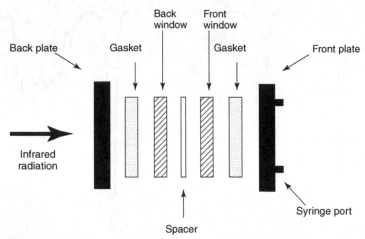

Figure 2.9 Schematic of a typical semi-permanent liquid cell. From Stuart, B., *Biological Applications of Infrared Spectroscopy*, ACOL Series, Wiley, Chichester, UK, 1997. © University of Greenwich, and reproduced by permission of the University of Greenwich.

types are filled by using a syringe and the syringe ports are sealed with PTFE plugs before sampling.

DQ 2.1
Which type of solution cell would you consider to be the easiest to maintain?

Answer
*The **demountable** is by far the easiest to maintain as it can be readily dismantled and cleaned. The windows can be repolished, a new spacer supplied and the cell reassembled. The **permanent** cells are difficult to clean and can become damaged by water. The pathlengths need to calibrated regularly if quantitative work is to be undertaken. Variable pathlength cells suffer from similar disadvantages and they are difficult to take apart. The calibration therefore suffers and the cells have to be calibrated regularly.*

An important consideration in the choice of infrared cells is the type of window material. The latter must be transparent to the incident infrared radiation and alkali halides are normally used in transmission methods. The cheapest material is sodium chloride (NaCl), but other commonly used materials are listed in Table 2.1.

Certain difficulties arise when using water as a solvent in infrared spectroscopy. The infrared modes of water are very intense and may overlap with the sample modes of interest. This problem may be overcome by substituting water with deuterium oxide (D_2O). The infrared modes of D_2O occur at different wavenumbers to those observed for water because of the mass dependence of

Table 2.1 Summary of some optical materials used in transmission infrared spectroscopy. From Stuart, B., *Modern Infrared Spectroscopy*, ACOL Series, Wiley, Chichester, UK, 1996. © University of Greenwich, and reproduced by permission of the University of Greenwich

Window material	Useful range (cm^{-1})	Refractive index	Properties
NaCl	40 000–600	1.5	Soluble in water; slightly soluble in alcohol; low cost; fair resistance to mechanical and thermal shock; easily polished
KBr	43 500–400	1.5	Soluble in water and alcohol; slightly soluble in ether; hygroscopic; good resistance to mechanical and thermal shock
CaF_2	77 000–900	1.4	Insoluble in water; resists most acids and bases; does not fog; useful for high-pressure work
BaF_2	66 666–800	1.5	Insoluble in water; soluble in acids and NH_4Cl; does not fog; sensitive to thermal and mechanical shock
KCl	33 000–400	1.5	Similar properties to NaCl but less soluble; hygroscopic
CsBr	42 000–250	1.7	Soluble in water and acids; hygroscopic
CsI	42 000–200	1.7	Soluble in water and alcohol; hygroscopic

the vibrational wavenumber. Table 2.2 lists the characteristic bands observed for both H_2O and D_2O. Where water is used as a solvent, NaCl cannot be employed as a infrared window material as it is soluble in water. Small path-lengths (~ 0.010 mm) are available in liquid cells and help reduce the intensities of the very strong infrared modes produced in the water spectrum. The small path-length also produces a small sample cavity, hence allowing samples in milligram quantities to be examined.

SAQ 2.2

What would be an appropriate material for liquid cell windows if an aqueous solution at pH 7 is to be examined?

Liquid films provide a quick method of examining liquid samples. A drop of liquid may be sandwiched between two infrared plates, which are then mounted in a cell holder.

Table 2.2 The major infrared bands of water and deuterium oxide. From Stuart, B., *Biological Applications of Infrared Spectroscopy*, ACOL Series, Wiley, Chichester, UK, 1997. © University of Greenwich, and reproduced by permission of the University of Greenwich

Wavenumber (cm^{-1})	Assignment
3920	O–H stretching
3490	O–H stretching
3280	O–H stretching
1645	H–O–H bending
2900	O–D stretching
2540	O–D stretching
2450	O–D stretching
1215	D–O–D bending

DQ 2.2

The method of liquid films is normally not used for volatile (with a boiling point less than 100°C) liquids. Why would this be necessary?

Answer

A common problem encountered in obtaining good quality spectra from liquid films is sample volatility. When the spectrum of a volatile sample is recorded, it progressively becomes weaker because evaporation takes place during the recording period. Liquids with boiling points below 100°C should be recorded in solution or in a short-pathlength sealed cell.

Before producing an infrared sample in solution, a suitable solvent must be chosen. In selecting a solvent for a sample, the following factors need to be considered: it has to dissolve the compound, it should be as non-polar as possible to minimize solute–solvent interactions, and it should not strongly absorb infrared radiation.

If quantitative analysis of a sample is required, it is necessary to use a cell of known pathlength. A guide to pathlength selection for different solution concentrations is shown in Table 2.3.

2.4.2 Solids

There are three general methods used for examining solid samples in transmission infrared spectroscopy; i.e. alkali halide discs, mulls and films. The choice of method depends very much on the nature of the sample to be examined. The use of alkali halide discs involves mixing a solid sample with a dry alkali halide powder. The mixture is usually ground with an agate mortar and pestle and

Table 2.3 Pathlength selection for solution cells. From Stuart, B., *Modern Infrared Spectroscopy*, ACOL Series, Wiley, Chichester, UK, 1996. © University of Greenwich, and reproduced by permission of the University of Greenwich

Concentration (%)	Pathlength (mm)
>10	0.05
1–10	0.1
0.1–1	0.2
< 0.1	>0.5

subjected to a pressure of about 10 ton in^{-2} (1.575×10^5 kg m^{-2}) in an evacuated die. This sinters the mixture and produces a clear transparent disc. The most commonly used alkali halide is potassium bromide (KBr), which is completely transparent in the mid-infrared region. Certain factors need to be considered when preparing alkali halide discs. The ratio of the sample to alkali halide is important; surprisingly little sample is needed and around 2 to 3 mg of sample with about 200 mg of halide is sufficient. The disc must not be too thick or too thin; thin discs are fragile and difficult to handle, while thick discs transmit too little radiation. A disc of about 1 cm diameter made from about 200 mg of material usually results in a good thickness of about 1 mm. If the crystal size of the sample is too large, excessive scattering of radiation results, particularly so at high wavenumbers (this is known as the *Christiansen effect*). The crystal size must be reduced, normally by grinding the solid using a mortar and pestle. If the alkali halide is not perfectly dry, bands due to water appear in the spectrum. Contributions due to water are difficult to avoid, and so the alkali halide should be kept desiccated and warm prior to use in order to minimize this effect.

The mull method for solid samples involves grinding the sample and then suspending this (about 50 mg) in one to two drops of a mulling agent. This is followed by further grinding until a smooth paste is obtained. The most commonly used mulling agent is Nujol (liquid paraffin), with its spectrum being shown in Figure 2.10. Although the mull method is quick and easy, there are some experimental factors to consider. The ratio of the sample to mulling agent must be correct. Too little sample, and there will be no sign of the sample in the spectrum. Too much sample and a thick paste will be produced and no radiation will be transmitted. A rough guide to mull preparation is to use a micro-spatula tip of sample to two to three drops of mulling agent. If the crystal size of the sample is too large, this leads to scattering of radiation, which gets worse at the high-wavenumber end of the spectrum. If the mull is not spread over the whole plate area, the beam of radiation

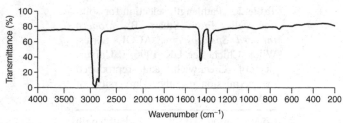

Figure 2.10 Infrared spectrum of Nujol (liquid paraffin). From Stuart, B., *Modern Infrared Spectroscopy*, ACOL Series, Wiley, Chichester, UK, 1996. © University of Greenwich, and reproduced by permission of the University of Greenwich.

passes part through the mull and only part through the plates, hence producing a distorted spectrum. The amount of sample placed between the infrared plates is an important factor; too little leads to a very weak spectrum showing only the strongest absorption bands. Too much mull leads to poor transmission of radiation so that the baseline may be at 50% or less.

It is sometimes possible to reduce the energy of a reference beam to a similar extent by use of an *attenuator*. Beam attenuators are placed in the sample compartment, working somewhat like a venetian blind, and the amount of radiation passing to the detector may be adjusted.

SAQ 2.3

The spectrum of a mull is shown in Figure 2.11. What is the problem with the mull produced and how would one go about remedying the problem?

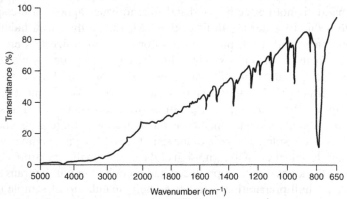

Figure 2.11 Infrared spectrum of a mull (cf. SAQ 2.3). From Stuart, B., *Modern Infrared Spectroscopy*, ACOL Series, Wiley, Chichester, UK, 1996. © University of Greenwich, and reproduced by permission of the University of Greenwich.

Films can be produced by either solvent casting or by melt casting. In solvent casting, the sample is dissolved in an appropriate solvent (the concentration depends on the required film thickness). A solvent needs to be chosen which not only dissolves the sample, but will also produce a uniform film. The solution is poured onto a levelled glass plate (such as a microscope slide) or a metal plate and spread to uniform thickness. The solvent may then be evaporated in an oven and, once dry, the film can be stripped from the plate. However, care must be taken as the heating of samples may cause degradation. Alternatively, it is possible to cast a film straight onto the infrared window to be used. Solid samples which melt at relatively low temperatures without decomposition can be prepared by melt casting. A film is prepared by 'hot-pressing' the sample in a hydraulic press between heated metal plates.

2.4.3 Gases

Gases have densities which are several orders of magnitude less than liquids, and hence pathlengths must be correspondingly greater, usually 10 cm or longer [2]. A typical gas cell is shown in Figure 2.12. The walls are of glass or brass, with the usual choice of windows. The cells can be filled by flushing or from a gas line. To analyse complex mixtures and trace impurities, longer pathlengths are necessary. As the sample compartment size in the instrument is limited, a multi-reflection gas cell is necessary to produce higher pathlengths. In such a cell, the infrared beam is deflected by a series of mirrors which reflect the beam back and forth many times until it exits the cell after having travelled the required

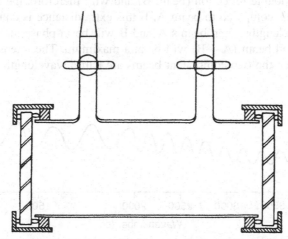

Figure 2.12 Schematic of a typical infrared gas cell. From Stuart, B., *Modern Infrared Spectroscopy*, ACOL Series, Wiley, Chichester, UK, 1996. © University of Greenwich, and reproduced by permission of the University of Greenwich.

equivalent pathlength. This type of cell allows a pathlength of up to 40 m to be attained.

2.4.4 Pathlength Calibration

When using transmission cells it can be useful to know precisely the pathlength of the cell, particularly for quantitative measurements. The cell pathlength can be measured by the method of counting interference fringes. If an empty cell with parallel windows is placed in the spectrometer and a wavelength range scanned, an interference pattern similar to that shown in Figure 2.13 will be obtained. The amplitude of the waveform will vary from 2 to 15%, depending on the state of the windows. The relationship between the pathlength of the cell, L, and the peak-to-peak fringes is given by the following:

$$L = \frac{n}{2(\bar{v}_1 - \bar{v}_2)} \tag{2.5}$$

where n is the number of complete peak-to-peak fringes between two maxima (or minima) at $\bar{v}_1$ and $\bar{v}_2$. If the spectrometer is calibrated in wavelengths, the n in Equation (2.5) can be converted to a more convenient form:

$$L = \frac{n(\lambda_1 \times \lambda_2)}{2(\lambda_1 - \lambda_2)} \tag{2.6}$$

where the values of the wavelengths, λ, are expressed in cm.

When a beam of radiation is directed into the face of a cell, most of the radiation will pass straight through (Figure 2.14, beam A). Some of the radiation will undergo a double reflection (beam B) and will, therefore, have travelled an extra distance $2L$ compared to beam A. If this extra distance is equal to a whole number of wavelengths, then beams A and B will be in-phase and the intensity of the transmitted beam (A + B) will be at a maximum. The intensity will be at a minimum when the two component beams are half a wavelength out-of-phase.

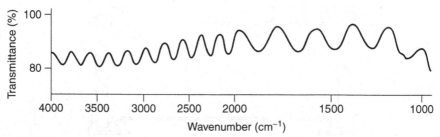

Figure 2.13 Interference pattern recorded with an empty cell in the sample beam. From Stuart, B., *Modern Infrared Spectroscopy*, ACOL Series, Wiley, Chichester, UK, 1996. © University of Greenwich, and reproduced by permission of the University of Greenwich.

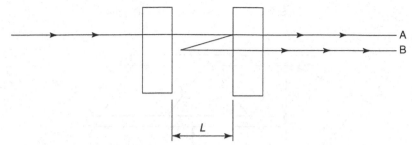

Figure 2.14 Beam of radiation passing through an empty cell.

SAQ 2.4

Using the interference pattern given in Figure 2.13, calculate the pathlength of the cell.

2.5 Reflectance Methods

Reflectance techniques may be used for samples that are difficult to analyse by the conventional transmittance methods. Reflectance methods can be divided into two categories. Internal reflectance measurements can be made by using an attenuated total reflectance cell in contact with the sample. There is also a variety of external reflectance measurements which involve an infrared beam reflected directly from the sample surface.

2.5.1 Attenuated Total Reflectance Spectroscopy

Attenuated total reflectance (ATR) spectroscopy utilizes the phenomenon of *total internal reflection* (Figure 2.15). A beam of radiation entering a crystal will undergo total internal reflection when the angle of incidence at the interface between the sample and crystal is greater than the critical angle, where the latter is a function of the refractive indices of the two surfaces. The beam penetrates a fraction of a wavelength beyond the reflecting surface and when a material that selectively absorbs radiation is in close contact with the reflecting surface, the beam loses energy at the wavelength where the material absorbs. The resultant attenuated radiation is measured and plotted as a function of wavelength by the spectrometer and gives rise to the absorption spectral characteristics of the sample.

The depth of penetration in ATR spectroscopy is a function of the wavelength, λ, the refractive index of the crystal, n_2, and the angle of incident radiation, θ. The

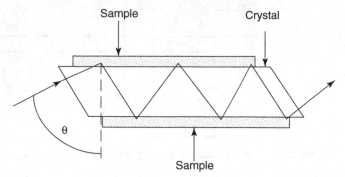

Figure 2.15 Schematic of a typical attenuated total reflectance cell. From Stuart, B., *Modern Infrared Spectroscopy*, ACOL Series, Wiley, Chichester, UK, 1996. © University of Greenwich, and reproduced by permission of the University of Greenwich.

depth of penetration, d_p, for a non-absorbing medium is given by the following:

$$d_p = (\lambda/n_1)/\left\{2\pi[\sin\theta - (n_1/n_2)^2]^{1/2}\right\} \qquad (2.7)$$

where n_1 is the refractive index of the sample.

The crystals used in ATR cells are made from materials that have low solubility in water and are of a very high refractive index. Such materials include zinc selenide (ZnSe), germanium (Ge) and thallium–iodide (KRS-5). The properties of these commonly used materials for ATR crystals are summarized in Table 2.4.

Different designs of ATR cells allow both liquid and solid samples to be examined. It is also possible to set up a flow-through ATR cell by including an inlet and outlet in the apparatus. This allows for the continuous flow of solutions through the cell and permits spectral changes to be monitored with

Table 2.4 Materials used as ATR crystals and their properties. From Stuart, B., *Modern Infrared Spectroscopy*, ACOL Series, Wiley, Chichester, UK, 1996. © University of Greenwich, and reproduced by permission of the University of Greenwich

Window material	Useful range (cm^{-1})	Refractive index	Properties
KRS-5 (thallium iodide)	17 000–250	2.4	Soluble in bases; slightly soluble in water; insoluble in acids; soft; highly toxic (handle with gloves)
ZnSe	20 000–500	2.4	Insoluble in water, organic solvents, dilute acids and bases
Ge	5000–550	4.0	Insoluble in water; very brittle

time. Multiple internal reflectance (MIR) and ATR are similar techniques, but MIR produces more intense spectra from multiple reflections. While a prism is usually used in ATR work, MIR uses specially shaped crystals that cause many internal reflections, typically 25 or more.

SAQ 2.5

The spectrum of a polymer film (refractive index, 1.5) was produced by using an ATR cell made of KRS-5 (refractive index, 2.4). If the incident radiation enters the cell crystal at an angle of 60°, what is the depth of penetration into the sample surface at:

(a) 1000 cm^{-1}

(b) 1500 cm^{-1}

(c) 3000 cm^{-1}?

2.5.2 Specular Reflectance Spectroscopy

In external reflectance, incident radiation is focused onto the sample and two forms of reflectance can occur, i.e. *specular* and *diffuse*. External reflectance measures the radiation reflected from a surface. The material must, therefore, be reflective or be attached to a reflective backing. A particularly appropriate application for this technique is the study of surfaces.

Specular reflectance occurs when the reflected angle of radiation equals the angle of incidence (Figure 2.16). The amount of light reflected depends on the angle of incidence, the refractive index, surface roughness and absorption properties of the sample. For most materials, the reflected energy is only 5–10%, but in regions of strong absorptions, the reflected intensity is greater. The resultant data appear different from normal transmission spectra, as 'derivative-like' bands result from the superposition of the normal extinction coefficient spectrum with the refractive index dispersion (based upon Fresnel's relationships). However, the reflectance spectrum can be corrected by using a *Kramers–Kronig transformation* (K–K transformation). The corrected spectrum then appears like the familiar transmission spectrum.

Increased pathlengths through thin coatings can be achieved by using grazing angles of incidence (up to 85°). Grazing angle sampling accessories allow measurements to be made on samples over a wide range of angles of incidence. Solid samples, particularly coatings on reflective surfaces, are simply placed on a flat surface. The technique is also commonly used for liquid samples that can be poured into a 'Teflon' trough. Oriented films on the liquid surface can be investigated by using this method.

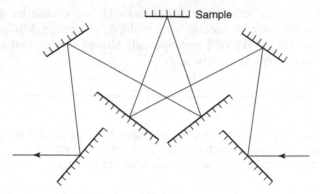

Figure 2.16 Illustration of specular reflectance.

2.5.3 Diffuse Reflectance Spectroscopy

In external reflectance, the energy that penetrates one or more particles is reflected in all directions and this component is called *diffuse reflectance*. In the diffuse reflectance (infrared) technique, commonly called DRIFT, a powdered sample is mixed with KBr powder. The DRIFT cell reflects radiation to the powder and collects the energy reflected back over a large angle. Diffusely scattered light can be collected directly from material in a sampling cup or, alternatively, from material collected by using an abrasive sampling pad. DRIFT is particularly useful for sampling powders or fibres. Figure 2.17 illustrates diffuse reflectance from the surface of a sample.

Kubelka and Munk developed a theory describing the diffuse reflectance process for powdered samples which relates the sample concentration to the scattered radiation intensity. The Kubelka–Munk equation is as follows:

$$(1 - R)^2/2R = c/k \tag{2.8}$$

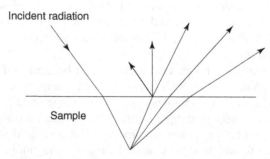

Figure 2.17 Illustration of diffuse reflectance.

where R is the absolute reflectance of the layer, c is the concentration and k is the molar absorption coefficient. An alternative relationship between the concentration and the reflected intensity is now widely used in near-infrared diffuse reflectance spectroscopy, namely:

$$\log (1/R) = k'c \qquad (2.9)$$

where k' is a constant.

2.5.4 *Photoacoustic Spectroscopy*

Photoacoustic spectroscopy (PAS) is a non-invasive reflectance technique with penetration depths in the range from microns down to several molecular mono-layers. Gaseous, liquid or solid samples can be measured by using PAS and the technique is particularly useful for highly absorbing samples. The photoacoustic effect occurs when intensity-modulated light is absorbed by the surface of a sample located in an acoustically isolated chamber filled with an inert gas. A spectrum is obtained by measuring the heat generated from the sample due to a re-absorption process. The sample absorbs photons of the modulated radiation,

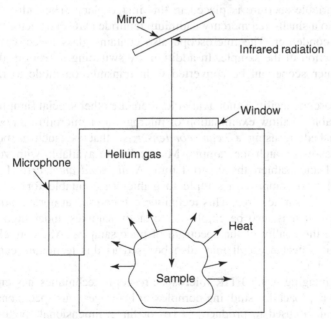

Figure 2.18 Schematic of a typical photoacoustic spectroscopy cell. From Stuart, B., *Modern Infrared Spectroscopy*, ACOL Series, Wiley, Chichester, UK, 1996. © University of Greenwich, and reproduced by permission of the University of Greenwich.

which have energies corresponding to the vibrational states of the molecules. The absorbed energy is released in the form of heat generated by the sample, which causes temperature fluctuations and, subsequently, periodic acoustic waves. A microphone detects the resulting pressure changes, which are then converted to electrical signals. Fourier-transformation of the resulting signal produces a characteristic infrared spectrum. Figure 2.18 shows a schematic diagram of a PAS cell.

2.6 Microsampling Methods

It is possible to combine an infrared spectrometer with a microscope facility in order to study very small samples [3–6]. In recent years, there have been considerable advances in FTIR microscopy, with samples of the order of microns being characterized. In FTIR microscopy, the microscope sits above the FTIR sampling compartment. Figure 2.19 illustrates the layout of a typical infrared microscope assembly. Infrared radiation from the spectrometer is focused onto a sample placed on a standard microscope $x-y$ stage. After passing through the sample, the infrared beam is collected by a Cassegrain objective which produces an image of the sample within the barrel of the microscope. A variable aperture is placed in this image plane. The radiation is then focused on to a small-area mercury cadmium telluride (MCT) detector by another Cassegrain condenser. The microscope also contains glass objectives to allow visual inspection of the sample. In addition, by switching mirrors in the optical train, the microscope can be converted from transmission mode to reflectance mode.

If a microscope facility is not available, there are other special sampling accessories available to allow examination of microgram or microlitre amounts. This is accomplished by using a *beam condenser* so that as much as possible of the beam passes through the sample. Microcells are available with volumes of around 4 μl and pathlengths up to 1 mm. A *diamond anvil cell* (DAC) uses two diamonds to compress a sample to a thickness suitable for measurement and increase the surface area. This technique can be used at normal atmospheric pressures, but it may also be employed to study samples under high pressures and improve the quality of the spectrum of trace samples. Alternatively, a multiple internal reflectance cell may also be used as this technique can produce stronger spectra.

Infrared imaging using FTIR microspectroscopic techniques has emerged as an effective approach for studying complex or heterogeneous specimens [7]. The technique can be used to produce a two- or three-dimensional 'picture' of the properties of a sample. This is achievable because, instead of reading the signal of

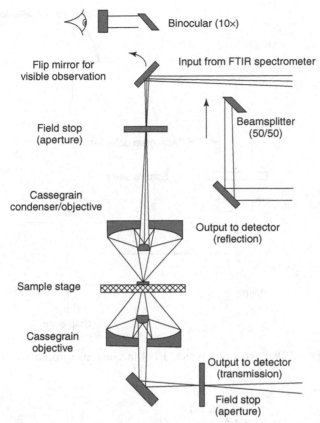

Figure 2.19 Layout of a typical FTIR microspectrometer. Reprinted from Katon, J. E., Sommer, A. J. and Lang, P. L., *Applied Spectroscopy Reviews*, Vol. 25, pp. 173–211 (1989–1990), courtesy of Marcel Dekker, Inc.

only one detector as in conventional FTIR spectroscopy, a large number of detector elements are read during the acquisition of spectra. This is possible due to the development of focal plane array (FPA) detectors. Currently, a step-scanning approach is used which means that that the moving mirror does not move continuously during data acquisition, but waits for each detector readout to be completed before moving on to the next position. This allows thousands of interferograms to be collected simultaneously and then transformed into infrared spectra.

Figure 2.20 illustrates the general layout of an FTIR imaging microspectrometer. The infrared beam from a Michelson interferometer is focused onto a sample with a reflective Cassegrain condenser. The light transmitted is collected by a Cassegrain objective and then focused onto an FPA detector. The imaging process

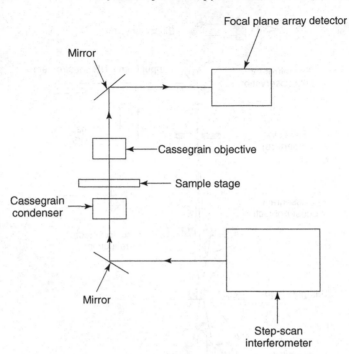

Figure 2.20 Layout of a typical FTIR imaging microspectrometer.

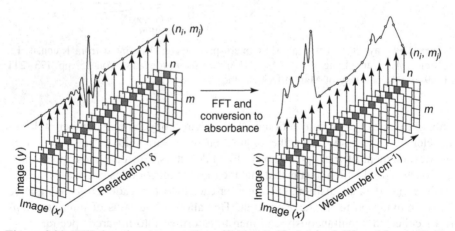

Figure 2.21 FTIR imaging data set. From Kidder, L. H., Levin, I. W. and Lewis, E. N., 'Infrared Spectroscopic Imaging Microscopy: Applications to Biological Systems', in *Proceedings of the 11th International Fourier Transform Spectroscopy Conference*, Athens, GA, USA, August 10–15, 1997, de Haseth, J. A. (Ed.), Figure 1, p. 148, American Institute of Physics, Melville, New York, 1998, pp. 148–158. Reproduced by permission of American Institute of Physics.

is illustrated in Figure 2.21. The data are collected as interferograms with each pixel on the array having a response determined by its corresponding location on the sample. Each point of the interferogram represents a particular moving mirror position and the spectral data are obtained by performing a Fourier-transform for each pixel on the array. Thus, each pixel (or spatial location) is represented by an infrared spectrum.

2.7 Chromatography–Infrared Spectroscopy

Infrared spectroscopy may be combined with each of a number of possible chromatographic techniques, with gas chromatography–infrared spectroscopy (GC–IR) being the most widely used [8, 9]. GC–IR allows for the identification of the components eluting from a gas chromatograph. In GC, the sample in a gaseous mobile phase is passed through a column containing a liquid or solid stationary phase. The retention of the sample depends on the degree of interaction with the stationary phase and its volatility: the higher the affinity of the sample for the stationary phase, then the more the sample partitions into that phase and the longer it takes before it passes through the chromatograph. The sample is introduced into the column, housed in an oven, via injection at one end and a detector monitors the effluent at the other end. A common method for coupling a gas chromatograph to an FTIR spectrometer is to use a light pipe, i.e. a heated flow cell which allows the continuous scanning of the effluent emerging from the GC column. Figure 2.22 shows a schematic diagram of a typical GC–IR system.

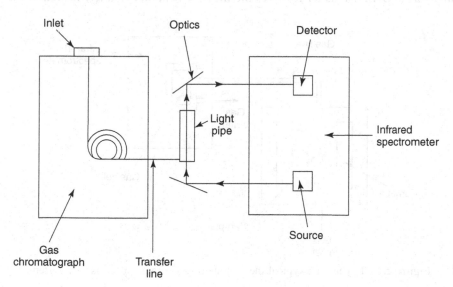

Figure 2.22 Layout of a typical GC–IR system.

The nature of this technique requires that interferograms are collected over short time intervals. Data can be displayed in real time and are commonly monitored as the changing spectrum of the GC effluent and the changing infrared absorption as a function of time. The latter is known as a *Gram–Schmidt chromatogram*.

Liquid chromatography (LC) may also be used in conjunction with infrared spectroscopy [10]. In this technique, the effluent from a liquid chromatograph is passed through a liquid flow-through cell. Supercritical fluid chromatography (SFC), where supercritical CO_2 is commonly used as a mobile phase, can be used with FTIR spectroscopy to improve detection limits.

2.8 Thermal Analysis–Infrared Spectroscopy

Infrared spectrometers may also be combined with thermal analysis instrumentation. Thermal analysis methods provide information about the temperature-dependent physical properties of materials. However, it is not always possible to gain information about the chemical changes associated with changes in temperature by using standard thermal analysis equipment. It is possible to combine thermal analysis apparatus with an infrared spectrometer in order to obtain a complete picture of the chemical and physical changes occurring in various thermal processes [11, 12].

The most common approach is to combine FTIR spectroscopy with a thermal method such as thermogravimetric analysis (TGA) to obtain an evolved gas analysis (EGA). The latter involves the measurement and characterization of the

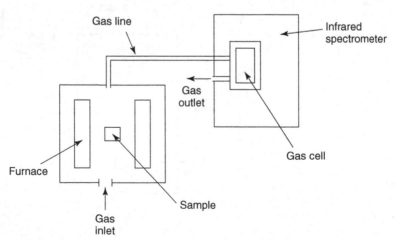

Figure 2.23 Layout of a typical thermal analysis–infrared spectroscopy system.

gaseous products which evolve from a sample when it is heated. Figure 2.23 presents a schematic layout of a typical thermal analysis–infrared spectroscopy system. In this, a sample is placed in a furnace while being suspended from a sensitive balance. The change in sample mass is recorded while the sample is maintained either at a required temperature or while being subjected to a programmed heating sequence. A TGA curve may be plotted as sample mass loss as a function of temperature or in a differential form where the change in sample mass with time is plotted as a function of temperature. The evolved gases can be carried from the furnace to the spectrometer where they can be examined in a long-pathlength gas cell. Data may be illustrated as a function of time by using a Gram–Schmidt plot.

2.9 Other Techniques

Variable-temperature cells, which are controlled to 0.1°C in the range −180 to 250°C, may be used in infrared spectrometers. An electrical heating system is used for temperatures above ambient, and liquid nitrogen with a heater for low temperatures. These cells can be used to study phase transitions and the kinetics of reactions. As well as transmission temperature cells, variable-temperature ATR cells and temperature cells for microsampling are available.

Infrared emission spectroscopy may be carried out by using a heated sample located in the emission port of the FTIR spectrometer as the radiation source.

Optical fibres may be used in conjunction with infrared spectrometers to carry out remote measurements. The fibres transfer the signal to and from a sensing probe and are made of materials that are flexible and 'infrared-transparent'.

For some samples, dipole moment changes may be in a fixed direction during a molecular vibration and, as such, can only be induced when the infrared radiation is polarized in that direction. Polarized infrared radiation can be produced by using a polarizer consisting of a fine grating of parallel metal wires. This approach is known as *linear infrared dichroism* [13].

SAQ 2.6

Which sampling technique would be the most appropriate in each of the cases listed below?

(a) A 1 g sample of white powder.

(b) A 10 μg sample of animal tissue.

(c) A powder containing a mixture of amphetamines.

Summary

Various aspects of the instrumentation used in infrared spectroscopy were dealt with in this chapter. Traditional dispersive spectrometers were described. The operation and capabilities of Fourier-transform infrared spectrometers were also discussed. Transmission methods for obtaining infrared spectra were also examined. The sampling methods which can be used for solids, solutions, liquids and gases were presented. The different reflectance methods that are now widely available, such as ATR spectroscopy, specular reflectance and diffuse reflectance, along with photoacoustic spectroscopy, were also introduced. The various microsampling techniques, which have emerged as effective methods for investigating small quantities of complex samples, were also described. Infrared spectrometers can also be used in conjunction with other analytical methods such as chromatography and thermal techniques and these were introduced in this chapter.

References

1. Griffiths, P. R. and de Haseth, J. A., *Fourier Transform Infrared Spectrometry*, Wiley, New York, 1986.
2. Günzler, H. and Gremlich, H.-U., *IR Spectroscopy: An Introduction*, Wiley-VCH, Weinheim, Germany, 2002.
3. Sommer, A. J., 'Mid-infrared Transmission Microspectroscopy', in *Handbook of Vibrational Spectroscopy*, Vol. 2, Chalmers, J. M. and Griffiths, P. R. (Eds), Wiley, Chichester, UK, 2002, pp. 1369–1385.
4. Katon, J. E., *Micron*, **27**, 303–314 (1996).
5. Humecki, H. J. (Ed.), *Practical Guide to Infrared Microspectroscopy*, Marcel Dekker, New York, 1999.
6. Messerschmidt, R. G. and Harthcock, M. A. (Eds), *Infrared Microspectroscopy: Theory and Applications*, Marcel Dekker, New York, 1998.
7. Kidder, L. H., Haka, A. S. and Lewis, E. N., 'Instrumentation for FT-IR Imaging', in *Handbook of Vibrational Spectroscopy*, Vol. 2, Chalmers, J. M. and Griffiths, P. R. (Eds), Wiley, Chichester, UK, 2002, pp. 1386–1404.
8. Visser, T., 'Gas Chromatography/Fourier Transform Infrared Spectroscopy', in *Handbook of Vibrational Spectroscopy*, Vol. 2, Chalmers, J. M. and Griffiths, P. R. (Eds), Wiley, Chichester, UK, 2002, pp. 1605–1626.
9. Ragunathan, N., Krock, K. A., Klawun, C., Sasaki, T. A. and Wilkins, C. L., *J. Chromatogr., A*, **856**, 349–397 (1999).
10. Somsen, G. W., Gooijer, C., Velthorst, N. H. and Brinkman, U. A. T., *J. Chromatogr., A*, **811**, 1–34 (1998).
11. Hellgeth, J. W., 'Thermal Analysis–IR Methods', in *Handbook of Vibrational Spectroscopy*, Vol. 2, Chalmers, J. M. and Griffiths, P. R. (Eds), Wiley, Chichester, UK, 2002, pp. 1699–1714.
12. Haines, P. J. (Ed.), *Principles of Thermal Analysis and Calorimetry*, The Royal Society of Chemistry, Cambridge, UK, 2002.
13. Buffeteau, T. and Pezolet, M., 'Linear Dichroism in Infrared Spectroscopy', in *Handbook of Vibrational Spectroscopy*, Vol. 1, Chalmers, J. M. and Griffiths, P. R. (Eds), Wiley, Chichester, UK, 2002, pp. 693–710.

Chapter 3
Spectral Analysis

Learning Objectives

- To recognize the characteristic bands that appear in the mid-infrared, near-infrared and far-infrared regions.
- To understand how hydrogen bonding affects an infrared spectrum.
- To develop a strategy for the interpretation of infrared spectra.
- To understand and use a variety of techniques to compensate for background absorption and overlapping peaks.
- To convert transmittance values to the corresponding absorbance values.
- To use the Beer–Lambert law for the quantitative analysis of samples studied using infrared spectroscopy.
- To analyse simple mixtures using their infrared spectra.
- To analyse multi-component systems using their infrared spectra.
- To recognize the calibration methods that may be applied to samples studied using infrared spectroscopy.

3.1 Introduction

Once an infrared spectrum has been recorded, the next stage of this experimental technique is interpretation. Fortunately, spectrum interpretation is simplified by the fact that the bands that appear can usually be assigned to particular parts of a molecule, producing what are known as *group frequencies*. The characteristic group frequencies observed in the mid-infrared region are discussed in this chapter. The types of molecular motions responsible for infrared bands in the near-infrared and far-infrared regions are also introduced.

Infrared Spectroscopy: Fundamentals and Applications B. Stuart
© 2004 John Wiley & Sons, Ltd ISBNs: 0-470-85427-8 (HB); 0-470-85428-6 (PB)

In Chapter 1, some of the factors which complicate the appearance of infrared spectra were defined, i.e. overtones and combination bands, coupling, Fermi resonance and vibration–rotation bands. Another factor, hydrogen bonding, may also contribute to notable changes in infrared spectra. The presence of this type of bonding in molecules can introduce additional, and sometimes misleading, information into the spectra. It is important to be aware of such factors before tackling the interpretation of a given spectrum.

Quantitative infrared spectroscopy can provide certain advantages over other analytical techniques. This approach may be used for the analysis of one component of a mixture, especially when the compounds in the mixture are alike chemically or have very similar physical properties (for example, structural isomers). In these instances, analysis using ultraviolet/visible spectroscopy, for instance, is difficult because the spectra of the components will be nearly identical. Chromatographic analysis may be of limited use because separation, of say isomers, is difficult to achieve. The infrared spectra of isomers are usually quite different in the *fingerprint* region. Another advantage of the infrared technique is that it can be non-destructive and requires a relatively small amount of sample.

In this present chapter, an introduction is provided on how infrared spectroscopy can be used for quantitative analysis. First, the various ways in which an infrared spectrum can be manipulated for analysis are outlined. Concentration is an important issue in quantitative analysis and the important relationships are introduced. Not only may quantitative infrared analysis be carried out on simple systems, it can also be applied to multi-component systems. Here, some straightforward examples will be used to demonstrate the analysis of the data obtained, while the calibration methods commonly applied to such infrared data will also be described.

3.2 Group Frequencies

3.2.1 Mid-Infrared Region

The mid-infrared spectrum (4000–400 cm^{-1}) can be approximately divided into four regions and the nature of a group frequency may generally be determined by the region in which it is located. The regions are generalized as follows: the X–H stretching region (4000–2500 cm^{-1}), the triple-bond region (2500–2000 cm^{-1}), the double-bond region (2000–1500 cm^{-1}) and the fingerprint region (1500–600 cm^{-1}).

The fundamental vibrations in the 4000–2500 cm^{-1} region are generally due to O–H, C–H and N–H stretching. O–H stretching produces a broad band that occurs in the range 3700–3600 cm^{-1}. By comparison, N–H stretching is usually observed between 3400 and 3300 cm^{-1}. This absorption is generally much sharper than O–H stretching and may, therefore, be differentiated. C–H stretching bands from aliphatic compounds occur in the range 3000–2850 cm^{-1}. If the C–H bond is adjacent to a double bond or aromatic ring, the C–H stretching wavenumber increases and absorbs between 3100 and 3000 cm^{-1}.

Triple-bond stretching absorptions fall in the 2500–2000 cm^{-1} region because of the high force constants of the bonds. C≡C bonds absorb between 2300 and 2050 cm^{-1}, while the nitrile group (C≡N) occurs between 2300 and 2200 cm^{-1}. These groups may be distinguished since C≡C stretching is normally very weak, while C≡N stretching is of medium intensity. These are the most common absorptions in this region, but you may come across some X–H stretching absorptions, where X is a more massive atom such as phosphorus or silicon. These absorptions usually occur near 2400 and 2200 cm^{-1}, respectively.

The principal bands in the 2000–1500 cm^{-1} region are due to C=C and C=O stretching. Carbonyl stretching is one of the easiest absorptions to recognize in an infrared spectrum. It is usually the most intense band in the spectrum, and depending on the type of C=O bond, occurs in the 1830–1650 cm^{-1} region. Note also that metal carbonyls may absorb above 2000 cm^{-1}. C=C stretching is much weaker and occurs at around 1650 cm^{-1}, but this band is often absent for symmetry or dipole moment reasons. C=N stretching also occurs in this region and is usually stronger.

It has been assumed so far that each band in an infrared spectrum can be assigned to a particular deformation of the molecule, the movement of a group of atoms, or the bending or stretching of a particular bond. This is possible for many bands, particularly stretching vibrations of multiple bonds that are 'well behaved'. However, many vibrations are not so well behaved and may vary by hundreds of wavenumbers, even for similar molecules. This applies to most bending and skeletal vibrations, which absorb in the 1500–650 cm^{-1} region, for which small steric or electronic effects in the molecule lead to large shifts. A spectrum of a molecule may have a hundred or more absorption bands present, but there is no need to assign the vast majority. The spectrum can be regarded as a 'fingerprint' of the molecule and so this region is referred to as the *fingerprint region*.

3.2.2 Near-Infrared Region

The absorptions observed in the near-infrared region (13 000–4000 cm^{-1}) are overtones or combinations of the fundamental stretching bands which occur in the 3000–1700 cm^{-1} region [1]. The bands involved are usually due to C–H, N–H or O–H stretching. The resulting bands in the near infrared are usually weak in intensity and the intensity generally decreases by a factor of 10 from one overtone to the next. The bands in the near infrared are often overlapped, making them less useful than the mid-infrared region for qualitative analysis. However, there are important differences between the near-infrared positions of different functional groups and these differences can often be exploited for quantitative analysis.

3.2.3 Far-Infrared Region

The far-infrared region is defined as the region between 400 and 100 cm^{-1} [2]. This region is more limited than the mid infrared for spectra–structure correlations, but does provide information regarding the vibrations of molecules containing heavy atoms, molecular skeleton vibrations, molecular torsions and crystal lattice vibrations. Intramolecular stretching modes involving heavy atoms can be helpful for characterizing compounds containing halogen atoms, organometallic compounds and inorganic compounds. Skeletal bending modes involving an entire molecule occur in the far infrared for molecules containing heavier atoms because bending modes are usually no more than one-half of the wavenumber of the corresponding stretching mode. Torsional modes arise because the rotation about single bonds is not 'free'. As a result, when certain small groups are bonded to a large group, they undergo a motion with respect to the heavier 'anchor' group. Crystal lattice vibrations are associated with the movement of whole molecular chains with respect to each other in crystalline solids.

3.3 Identification

There are a few general rules that can be used when using a mid-infrared spectrum for the determination of a molecular structure. The following is a suggested strategy for spectrum interpretation:

1. Look first at the high-wavenumber end of the spectrum (>1500 cm^{-1}) and concentrate initially on the major bands.

2. For each band, 'short-list' the possibilities by using a correlation chart.

3. Use the lower-wavenumber end of the spectrum for the confirmation or elaboration of possible structural elements.

4. Do not expect to be able to assign every band in the spectrum.

5. Keep 'cross-checking' wherever possible. For example, an aldehyde should absorb near 1730 cm^{-1} and in the region 2900–2700 cm^{-1} (see Chapter 4 for such band assignments).

6. Exploit negative evidence as well as positive evidence. For example, if there is no band in the 1850–1600 cm^{-1} region, it is most unlikely that a carbonyl group is present.

7. Band intensities should be treated with some caution. Under certain circumstances, they may vary considerably for the same group.

8. Take care when using small wavenumber changes. These can be influenced by whether the spectrum was run as a solid or liquid, or in solution. If in solution, some bands are very 'solvent-sensitive'.

9. Do not forget to subtract solvent bands if possible – these could be confused with bands from the sample.

It is not always possible by examination of the infrared spectrum of a compound alone to identify it unequivocally. It is normal to use infrared spectroscopy in conjunction with other techniques, such as chromatographic methods, mass spectrometry, NMR spectroscopy and various other spectroscopic techniques.

Advances in computer retrieval techniques have extended the range of information available from an infrared instrument by allowing comparison of an unknown spectrum with a 'bank' of known compounds. This was only possible in the past by manually searching libraries of spectra. Once the compound was classified, the atlas of spectra could be searched for an identical or very similar spectrum. Computers have accelerated this search process. The programs work by hunting through stored data to match intensities and wavenumbers of absorption bands [3]. The computer will then output the best fits and the names of the compounds found.

3.4 Hydrogen Bonding

The presence of hydrogen bonding is of great importance in a range of molecules. For instance, the biological activity of deoxyribonucleic acid (DNA) relies on this type of bonding. Hydrogen bonding is defined as the attraction that occurs between a highly electronegative atom carrying a non-bonded electron pair (such as fluorine, oxygen or nitrogen) and a hydrogen atom, itself bonded to a small highly electronegative atom. An example of this type of bonding is illustrated by the interactions between water molecules in Figure 3.1. This is an example of *intermolecular* hydrogen bonding. It is also possible for a hydrogen bond to form between appropriate groups within the one molecule. This is known as *intramolecular* hydrogen bonding and is illustrated by the protein structure shown in Figure 3.2. The segments shown belong to the same protein molecule.

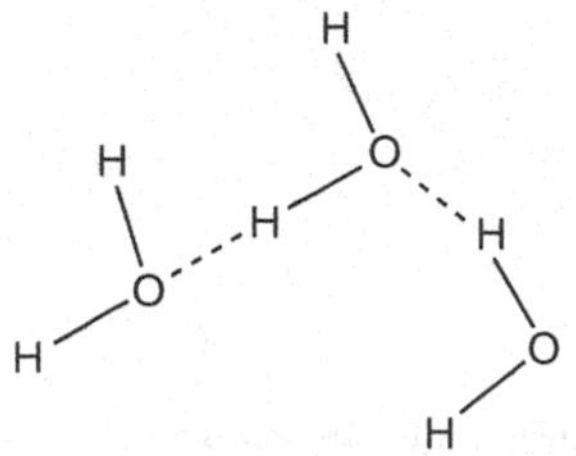

Figure 3.1 Hydrogen bonding of water molecules.

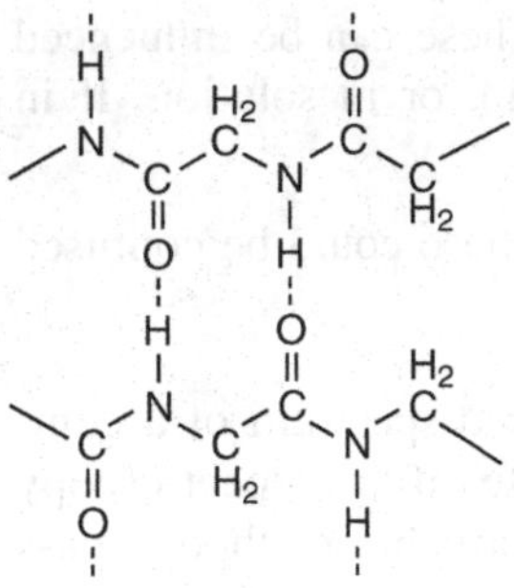

Figure 3.2 Intramolecular hydrogen bonding in a protein.

DQ 3.1

Would chloroform be expected to form hydrogen bonds? If so, would those bonds be intramolecular or intermolecular?

Answer

Chloroform ($CHCl_3$) possesses a donor hydrogen but no suitable lone pair exists in the molecule to form a strong hydrogen bond. A very weak hydrogen bond could form with the chlorine lone pairs. Thus, this is a marginal case of intermolecular hydrogen bonding.

SAQ 3.1

Would the O–H stretching mode be expected to vary in intensity when hydrogen bonded. If so, would it decrease or increase?

Hydrogen bonding is a very important effect in infrared spectroscopy. This bonding influences the bond stiffness and so alters the frequency of vibration. For example, for a hydrogen bond in an alcohol, the O–H stretching vibration in a hydrogen-bonded dimer is observed in the $3500–2500$ cm^{-1} range, rather than in the usual $3700–3600$ cm^{-1} range (Figure 3.3).

Many solvents are capable of forming hydrogen bonds to solutes. Three classes of solvent exist which may lead to trouble when used for hydrogen-bonding

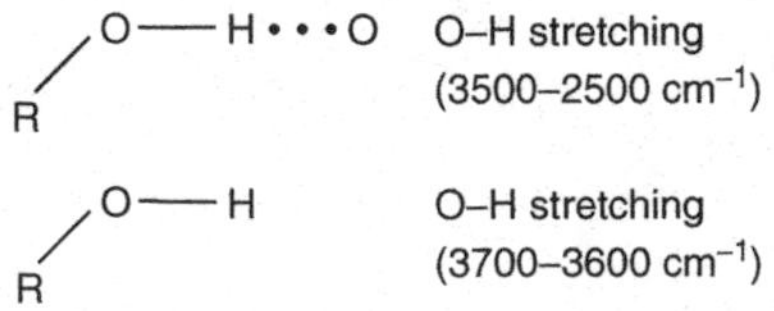

Figure 3.3 Effect of hydrogen bonding on an O–H stretching vibration.

studies is solution. First, compounds which contain hydrogen donor groups, e.g. halogenated compounds which contain a sufficient number of halogens to activate the hydrogens present, such as chloroform. Secondly, compounds which contain non-bonded electron pairs, such as ethers, aldehydes and tertiary amines. In addition, there are compounds which contain both types of group, e.g. water and alcohols. There are only a few solvents that do not have the above characteristics, such as carbon tetrachloride (CCl_4) and carbon disulfide (CS_2). These still contain lone electron pairs, but being on S and Cl are less available, and any interactions will be extremely weak.

Apart from solvent effects, concentration and temperature also affect the degree of hydrogen bonding in a compound. The lower the concentration, then the less chance there is of two molecules colliding. It follows that the degree of hydrogen bonding decreases with decreasing concentration.

DQ 3.2

What would be the effect of an increase in temperature on the infrared spectrum of a hydrogen-bonded compound?

Answer

Increasing temperature means that each molecule will have more energy on average and weak associative forces, such as hydrogen bonds, are likely to be broken. This should lead to a lesser degree of hydrogen bonding, and thus changes in wavenumber to greater values would be observed for groups forming the hydrogen bond.

3.5 Spectrum Manipulation

There are a number of techniques available to users of infrared spectrometers that help with both the qualitative and quantitative interpretation of spectra.

3.5.1 Baseline Correction

It is usual in quantitative infrared spectroscopy to use a baseline joining the points of lowest absorbance on a peak, preferably in reproducibly flat parts of the absorption line. The absorbance difference between the baseline and the top of the band is then used. An example of a baseline construction is shown in Figure 3.4.

3.5.2 Smoothing

'Noise' in a spectrum can be diminished by a smoothing process. After a spectrum is smoothed, it becomes similar to the result of an experiment at a lower

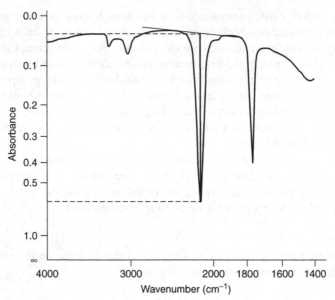

Figure 3.4 An example of a baseline correction. From Stuart, B., *Modern Infrared Spectroscopy*, ACOL Series, Wiley, Chichester, UK, 1996. © University of Greenwich, and reproduced by permission of the University of Greenwich.

resolution. The features are blended into each other and the noise level decreases. A smoothing function is basically a convolution between the spectrum and a vector whose points are determined by the degree of smoothing applied. Generally, a degradation factor is required, which will be some positive integer. A low value, say one, will produce only subtle changes, while a high value has a more pronounced effect on the spectrum.

3.5.3 Difference Spectra

The most straightforward method of analysis for complex spectra is *difference spectroscopy*. This technique may be carried out by simply subtracting the infrared spectrum of one component of the system from the combined spectrum to leave the spectrum of the other component. If the interaction between components results in a change in the spectral properties of either one or both of the components, the changes will be observed in the difference spectra. Such changes may manifest themselves via the appearance of positive or negative peaks in the spectrum.

Spectral subtraction may be applied to numerous applications and can be used for the data collected for solutions. In order to obtain the spectrum of a solution, it is necessary to record spectra of both the solution and the solvent alone. The solvent spectrum may then be subtracted from the solution spectrum. Figure 3.5,

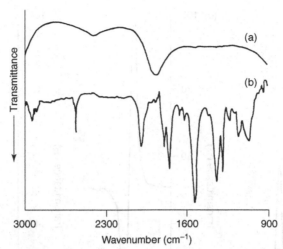

Figure 3.5 Infrared spectra of (a) a 1% (wt/vol) solution of aspirin in water and (b) the same solution after subtraction of the water spectrum. From Stuart, B., *Modern Infrared Spectroscopy*, ACOL Series, Wiley, Chichester, UK, 1996. © University of Greenwich, and reproduced by permission of the University of Greenwich.

where the strong infrared spectrum of water has been removed from a relatively weak spectrum of a 1% (wt/vol) solution of aspirin, illustrates how subtraction can be very useful in FTIR spectroscopy. However, care must be exercised because the concentration of the solvent alone is greater than that of the solvent in the solution and negative peaks may appear in the regions of solvent absorption.

Under certain circumstances, the spectrum due to the solvent may be very intense, making simple subtraction impossible. This situation can make it difficult to investigate the sample spectrum, as solvent bands may overlap with the region under investigation. In such experiments, attenuated total reflectance (ATR) spectroscopy provides a suitable approach. The nature of the ATR technique produces a less intense solvent contribution to the overall infrared spectrum and so solvent spectra can be more readily subtracted from the sample spectrum of interest by using this method.

Subtraction may also be applied to solid samples and is especially useful for mulls. The spectrum of the mulling agent can be subtracted, hence giving the spectrum of the solid only.

3.5.4 Derivatives

Spectra may also be differentiated. Figure 3.6 shows a single absorption peak and its first and second derivative. The benefits of derivative techniques are twofold. Resolution is enhanced in the first derivative since changes in the gradient are examined. The second derivative gives a negative peak for each band and shoulder in the absorption spectrum.

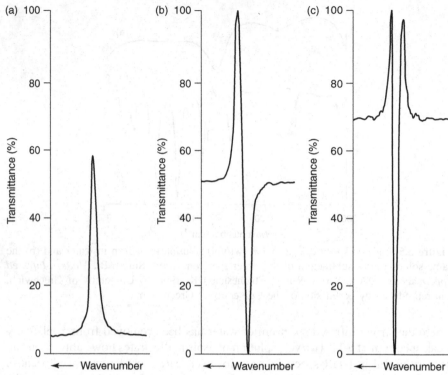

Figure 3.6 Differentiation of spectra: (a) single absorption peak; (b) first derivative; (c) second derivative. From Stuart, B., *Modern Infrared Spectroscopy*, ACOL Series, Wiley, Chichester, UK, 1996. © University of Greenwich, and reproduced by permission of the University of Greenwich.

The advantage of derivatization is more readily appreciated for more complex spectra and Figure 3.7 shows how differentiation may be used to resolve and locate peaks in an 'envelope'. Note that sharp bands are enhanced at the expense of broad ones and this may allow for the selection of a suitable peak, even when there is a broad band beneath.

With FTIR spectrometers, it is possible to apply what is known as *Fourier-derivation*. During this process, the spectrum is first transformed into an interferogram. It is then multiplied by an appropriate weighting function and finally it is 're-transformed' to give the derivative. This technique provides more sensitivity than conventional derivatization.

3.5.5 Deconvolution

Deconvolution is the process of compensating for the intrinsic linewidths of bands in order to resolve overlapping bands [4]. This technique yields spectra that have

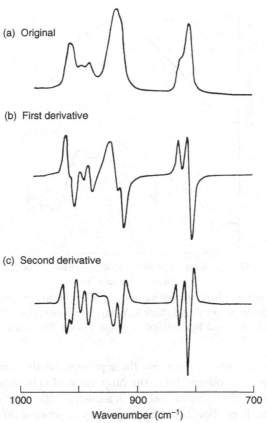

(a) Original

(b) First derivative

(c) Second derivative

1000   900   700

Wavenumber (cm^{-1})

Figure 3.7 Complex absorption band (a), plus corresponding first (b) and second (c) derivatives. From *Perkin-Elmer Application Notes on Derivative Spectroscopy*, Perkin-Elmer, Norwalk, CT, USA, 1984, and reproduced with permission of the Perkin-Elmer Corporation.

much narrower bands and is able to distinguish closely spaced features. The instrumental resolution is not increased, but the ability to differentiate spectral features can be significantly improved. This is illustrated by Figure 3.8, which shows a broad band before and after deconvolution has been applied. The peaks at quite close wavenumbers are now easily distinguished.

The deconvolution technique generally involves several steps: computation of an interferogram of the sample by computing the inverse Fourier-transform of the spectrum, multiplication of the interferogram by a smoothing function and by a function consisting of a Gaussian–Lorentzian bandshape, and Fourier-transformation of the modified interferogram.

The deconvolution procedure is typically repeated iteratively for best results. At iteration, the lineshape is adjusted in an attempt to provide narrower bands

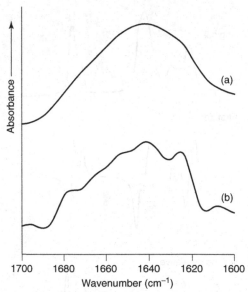

Figure 3.8 Broad infrared band (a) before and (b) after deconvolution. From Stuart, B., *Modern Infrared Spectroscopy*, ACOL Series, Wiley, Chichester, UK, 1996. © University of Greenwich, and reproduced by permission of the University of Greenwich.

without excessive distortion. There are three parameters that can be adjusted to 'tune' the lineshape, as follows. First, the proportion of Gaussian and Lorentzian lineshapes: the scale of these components is adjusted depending on the predicted origins of the bandshape. For instance, it will vary depending on whether a solid, liquid or gas is being investigated. The second factor is the half-width: this is the width of the lineshape. The half-width is normally the same as or larger than the intrinsic linewidth (full-width at half-height) of the band. If the value specified is too small, then the spectrum tends to show only small variations in intensity. If too large, distinctive negative side-lobes are produced. The narrowing function is a further factor: this is the degree of narrowing attempted on a scale from 0 to 1. If the narrowing function is specified too small, then the resulting spectrum will not be significantly different from the original, while if too large, the spectrum may produce false peaks as noise starts to be deconvolved.

3.5.6 Curve-Fitting

Quantitative values for band areas of heavily overlapped bands can be achieved by using curve-fitting procedures. Many of these are based on a least-squares minimization procedure. Least-squares curve-fitting covers the general class of techniques whereby one attempts to minimize the sum of the squares of the difference between an experimental spectrum and a computed spectrum generated by

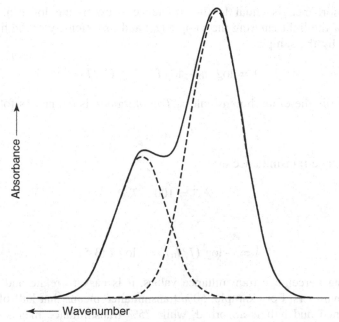

Figure 3.9 Illustration of curve-fitting of overlapping infrared bands.

summing the component curves. Generally, the procedure involves entering the values of the wavenumbers of the component bands (determined by using derivatives and/or deconvolution) and then the program determines the best estimate of the parameters of the component curves.

Apart from the obvious variables of peak height and width, the type of bandshape needs to be considered. The class of bandshape of an infrared spectrum depends on the type of sample. A choice of Gaussian, Lorentzian or a combination of these band shapes is usually applied. Figure 3.9 illustrates the results of a curve-fitting process.

3.6 Concentration

The Beer–Lambert law is used to relate the amount of light transmitted by a sample to the thickness of the sample. The *absorbance* of a solution is directly proportional to the thickness and the concentration of the sample, as follows:

$$A = \varepsilon c l \qquad (3.1)$$

where A is the absorbance of the solution, c the concentration and l the pathlength of the sample. The constant of proportionality is usually given the Greek symbol epsilon, ε, and is referred to as the *molar absorptivity*.

The absorbance is equal to the difference between the logarithms of the intensity of the light entering the sample (I_0) and the intensity of the light transmitted (I) by the sample:

$$A = \log\ I_0 - \log I = \log\ (I_0/I) \tag{3.2}$$

Absorbance is therefore dimensionless. *Transmittance* is defined as follows:

$$T = I/I_0 \tag{3.3}$$

and percentage transmittance as:

$$\%T = 100 \times T \tag{3.4}$$

Thus:

$$A = -\log\ (I/I_0) = -\log T \tag{3.5}$$

When using percentage transmittance values, it is easy to relate and to understand the numbers. For example, 50% transmittance means that half of the light is transmitted and half is absorbed, while 75% transmittance means that three quarters of the light is transmitted and one quarter absorbed.

DQ 3.3

What would be the absorbance of a solution that has a percentage transmittance of: (a) 100%; (b) 50%; (c) 10%; (d) 0%?

Answer

The use of Equations (3.4) and (3.5) gives:

(a) $A = -log\ (100/100) = 0;$

(b) $A = -log\ (50/100) = 0.303;$

(c) $A = -log\ (10/100) = 1.0;$

(d) $A = -log\ (0/100) = infinity.$

The important point to gain from these figures is that at an absorbance of 1.0, 90% of the light is being absorbed, and so the instrument's detector does not have much radiation with which to work.

SAQ 3.2

A 1.0% (wt/vol) solution of hexanol has an infrared absorbance of 0.37 at 3660 cm^{-1} in a 1.0 mm cell. Calculate its molar absorptivity at this wavenumber.

The Beer–Lambert law tells us that a plot of absorbance against concentration should be linear with a gradient of εl and will pass through the origin. In theory, to analyse a solution of unknown concentration, solutions of known concentration need to be prepared, a suitable band chosen, the absorbance at this wavenumber measured, and a calibration graph plotted. The concentration o the compound in solution can be read from the calibration graph, given that its absorbance is known.

There are a few factors which need to be considered when taking this approach. First, the preparation of solutions of known concentrations: these concentrations have to give sensible absorbance values – not too weak and not too intense. There is also the choice of a suitable absorption peak: this technique needs to be as sensitive as possible and so an intense peak should be chosen. However, often infrared spectra have many, sometimes overlapping peaks. A peak isolated from others, with a high molar absorptivity, should be found. A further problem that sometimes arises, especially in the spectra of solid samples, is the presence of asymmetric bands. In such cases, peak height cannot be used because the baseline will vary from sample to sample and peak area measurements must be used instead. FTIR spectrometers have accompanying software that can carry out these calculations (calibration methods are introduced below in Section 3.9). Quantitative measurements need to be carried out on absorbance spectra. Thus, transmittance spectra need to be converted to absorbance spectra.

The determination of the concentration of a gas needs to be considered separately from solids and liquids [5]. The ideal gas law relates the physical properties of a gas to its concentration. This law states that:

$$PV = nRT \tag{3.6}$$

where P is the pressure, V the volume, n the number of moles, R the universal gas constant and T the temperature. Given that the concentration of a gas in an infrared cell is given by $c = n/V$, Equation (3.1) may be rewritten as follows:

$$A = P\varepsilon l/RT \tag{3.7}$$

Thus, for gases the absorbance depends also upon the pressure and the temperature of the gas.

3.7 Simple Quantitative Analysis

3.7.1 Analysis of Liquid Samples

The quantitative analysis of a component in solution can be successfully carried out given that there is a suitable band in the spectrum of the component of interest. The band chosen for analysis should have a high molar absorptivity, not overlap with other peaks from other components in the mixture or the solvent, be symmetrical, and give a linear calibration plot of absorbance versus concentration.

Most simple quantitative infrared methods of analysis use the intensities of the C=O, N–H or O–H groups. The C=O stretching band is the most commonly used because it is a strong band in a spectral region relatively free of absorption by other functional groups. In addition, the carbonyl band is not as susceptible as the O–H and N–H bands to chemical change or hydrogen bonding.

Simple quantitative analysis can be illustrated by using the example of aspirin dissolved in chloroform. The best peak to choose in this example is the C=O stretching band of aspirin, observed at 1764 cm^{-1}, because it is an intense peak and

Table 3.1 Calibration data for aspirin solutions. From Stuart, B., *Biological Applications of Infrared Spectroscopy*, ACOL Series, Wiley, Chichester, UK, 1997. © University of Greenwich, and reproduced by permission of the University of Greenwich

Concentration (mg ml^{-1})	Absorbance at 1764 cm^{-1}
0	0.000
25	0.158
50	0.285
75	0.398
100	0.501

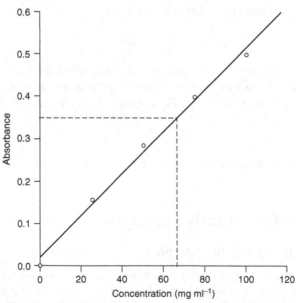

Figure 3.10 Calibration graph for aspirin solutions. From Stuart, B., *Biological Applications of Infrared Spectroscopy*, ACOL Series, Wiley, Chichester, UK, 1997. © University of Greenwich, and reproduced by permission of the University of Greenwich.

lies in a region where there is no interference from the chloroform spectrum. The next step is to draw a calibration plot of absorbance versus aspirin concentration. A series of chloroform solutions of known aspirin concentrations is prepared and the infrared spectrum of each solution recorded by using a 0.1 mm NaCl transmission cell. The absorbance of the 1764 cm^{-1} band is measured for each solution, with the results obtained being listed in Table 3.1. These data can then be graphed to produce a linear plot, as shown in Figure 3.10. The next stage in the analysis is to determine the amount of aspirin present in a chloroform solution of unknown concentration. The infrared spectrum of the unknown sample is recorded and the absorbance of the 1764 cm^{-1} carbonyl band was measured as 0.351. Using the calibration graph, this corresponds to a concentration of 67 mg ml^{-1}.

SAQ 3.3

A solution of caffeine in chloroform is provided for analysis, with the concentration of caffeine in the solution being required. Caffeine in chloroform shows a distinct carbonyl band at 1656 cm^{-1} in the infrared spectrum (recorded by using a 0.1 mm NaCl cell). The infrared information provided by standard solutions of caffeine in chloroform is listed below in Table 3.2. Given that the unknown sample produces an absorbance of 0.166 at 1656 cm^{-1} in its infrared spectrum, determine the concentration of this caffeine sample. Assume that there is no interference from other components in the sample.

Table 3.2 Calibration data for caffeine solutions (cf. SAQ 3.3). From Stuart, B., *Biological Applications of Infrared Spectroscopy*, ACOL Series, Wiley, Chichester, UK, 1997. © University of Greenwich, and reproduced by permission of the University of Greenwich

Concentration (mg ml^{-1})	Absorbance at 1656 cm^{-1}
0	0.000
5	0.105
10	0.190
15	0.265
20	0.333

SAQ 3.4

Commercial propanol often contains traces of acetone formed by oxidation. Acetone shows a distinct band at 1719 cm^{-1} suitable for quantitative analysis. The absorbance values at 1719 cm^{-1} versus the concentrations of a series of prepared solutions containing known quantities of acetone are listed below in Table 3.3.

The density of acetone is 0.790 g cm^{-3} and the pathlength used was 0.1 mm. The infrared spectrum of a 10 vol% solution of commercial propanol in CCl_4 in a 0.1 mm pathlength cell was recorded and a band of absorbance 0.337 at 1719 cm^{-1} is observed in the spectrum. Determine the concentration (in mol l^{-1}) of acetone in this solution and calculate the percentage acetone in the propanol.

Table 3.3 Calibration data for solutions of acetone in propanol (cf. SAQ 3.4). From Stuart, B., *Modern Infrared Spectroscopy*, ACOL Series, Wiley, Chichester, UK, 1996. © University of Greenwich, and reproduced by permission of the University of Greenwich

Acetone in CCl_4 (vol%)	Absorbance at 1719 cm^{-1}
0.25	0.123
0.50	0.225
1.00	0.510
1.50	0.736
2.00	0.940

Infrared spectroscopy can be used to measure the number of functional groups in a molecule, for example, the number of –OH or –NH$_2$ groups. It has been found that the molar absorptivities of the bands corresponding to the group are proportional to the number of groups that are present, that is, each group has its own intensity which does not vary drastically from molecule to molecule. This approach has been used to measure chain lengths in hydrocarbons by using the C–H deformation, the methylene group at 1467 and 1305 cm^{-1}, and the number of methyl groups in polyethylene.

3.7.2 Analysis of Solid Samples

Simple solid mixtures may also be quantitatively analysed. These are more susceptible to errors because of the scattering of radiation. Such analyses are usually carried out with KBr discs or in mulls. The problem here is the difficulty in measuring the pathlength. However, this measurement becomes unnecessary when an internal standard is used. When using this approach, addition of a constant known amount of an internal standard is made to all samples and calibration standards. The calibration curve is then obtained by plotting the ratio of the absorbance of the analyte to that of the internal standard, against the concentration of the analyte. The absorbance of the internal standard varies linearly with the sample thickness and thus compensates for this parameter. The discs or mulls must be prepared under exactly the same conditions to avoid intensity changes or shifts in band positions.

The standard must be carefully chosen and it should ideally possess the following properties: have a simple spectrum with very few bands; be stable to heat and not absorb moisture; be easily reduced to a particle size less than the incident radiation without lattice deformation; be non-toxic, giving clear discs in a short time; be readily available in the pure state. Some common standards used include calcium carbonate, sodium azide, napthalene and lead thiocyanate.

3.8 Multi-Component Analysis

The analysis of a component in a complex mixture presents special problems. In this section, the approach to determining the concentrations of a number of components in a mixture will be examined. The simple approach taken here serves to illustrate the fundamental ideas of analysis of mixtures, while the mathematical methods that may be employed to analyse multicomponent infrared data are summarized below in Section 3.9.

The quantitative analysis of a multi-component system is illustrated by its application to the simultaneous determination of a mixture of xylene isomers. Commercial xylene is a mixture of isomers, i.e. 1,2-dimethylbenzene (*o*-xylene), 1,3-dimethylbenzene (*m*-xylene) and 1,4-dimethylbenzene (*p*-xylene). The

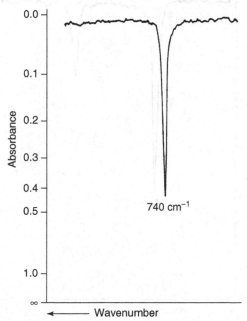

Figure 3.11 Infrared spectrum of *o*-xylene in cyclohexane (1 vol%, 0.1 mm pathlength). From Stuart, B., *Modern Infrared Spectroscopy*, ACOL Series, Wiley, Chichester, UK, 1996. © University of Greenwich, and reproduced by permission of the University of Greenwich.

spectra of these three *pure* xylenes in cyclohexane solutions (Figures 3.11, 3.12 and 3.13) all show strong bands in the $800-600$ cm^{-1} region. Cyclohexane has a very low absorbance in this region and is therefore a suitable solvent for the analysis. The infrared spectrum of a commercial sample of xylene is given in Figure 3.14.

The concentration of the three isomers may be estimated in the commercial sample. First, the absorbances of the xylenes at 740, 770 and 800 cm^{-1} from the standards shown in Figures 3.11, 3.12 and 3.13 need to be measured, as follows:

$$o\text{-xylene} \quad 740 \text{ cm}^{-1} \quad A = 0.440 - 0.012 = 0.428$$

$$m\text{-xylene} \quad 770 \text{ cm}^{-1} \quad A = \frac{0.460 - 0.015}{2} = 0.223$$

$$p\text{-xylene} \quad 800 \text{ cm}^{-1} \quad A = \frac{0.545 - 0.015}{2} = 0.265$$

The values for *m*-xylene and *p*-xylene are divided by two as these solutions are twice as concentrated. The absorbance values have also been corrected for a non-zero baseline in the region. The absorbance values are proportional to the molar

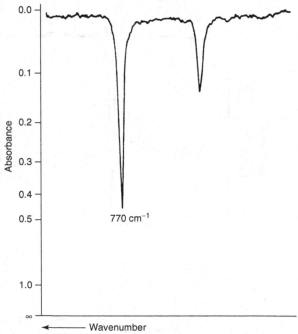

Figure 3.12 Infrared spectrum of *m*-xylene in cyclohexane (2 vol%, 0.1 mm pathlength). From Stuart, B., *Modern Infrared Spectroscopy*, ACOL Series, Wiley, Chichester, UK, 1996. © University of Greenwich, and reproduced by permission of the University of Greenwich.

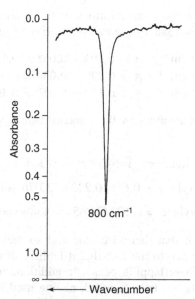

Figure 3.13 Infrared spectrum of *p*-xylene in cyclohexane (2 vol%, 0.1 mm pathlength). From Stuart, B., *Modern Infrared Spectroscopy*, ACOL Series, Wiley, Chichester, UK, 1996. © University of Greenwich, and reproduced by permission of the University of Greenwich.

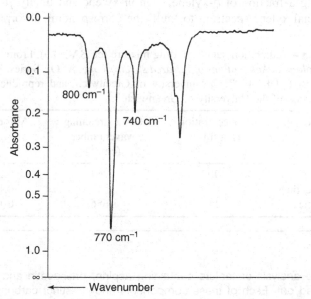

Figure 3.14 Infrared spectrum of commercial xylene in cyclohexane (5 vol%, 0.1 mm pathlength). From Stuart, B., *Modern Infrared Spectroscopy*, ACOL Series, Wiley, Chichester, UK, 1996. © University of Greenwich, and reproduced by permission of the University of Greenwich.

absorptivities, and hence the concentrations of the xylenes can be estimated in the mixture once the absorbances are measured. From Figure 3.14, we obtain:

$$740 \text{ cm}^{-1} \quad A = 0.194 - 0.038 = 0.156$$
$$770 \text{ cm}^{-1} \quad A = 0.720 - 0.034 = 0.686$$
$$800 \text{ cm}^{-1} \quad A = 0.133 - 0.030 = 0.103$$

Dividing these absorbance values by the standard values, gives the vol% of each isomer in the mixture:

$$o\text{-xylene} = 0.156/0.428 = 0.364 \text{ vol}\%$$

$$m\text{-xylene} = 0.686/0.223 = 3.076 \text{ vol}\%$$

$$p\text{-xylene} = 0.103/0.265 = 0.389 \text{ vol}\%$$

It should be pointed out that there are potential sources of error in these values. There can be error due to the fact that it is difficult to select a baseline for the analysis because of overlapping peaks. In addition, any non-linearity of the Beer–Lambert law plots has been ignored, having used one value to determine the constant of proportionality above for each solution. Alternatively, this same problem for xylene can be determined by spectral subtraction. If a sample of the impurity is available, mixtures of the starting material and product can be separated by subtraction of the spectra of the mixture and impurities. This is achieved by subtracting a fraction of o-xylene, then m-xylene and finally p-xylene from the commercial xylene spectrum to 'null' the corresponding absorptions.

Table 3.4 Calibration data for a drug mixture (cf. SAQ 3.5). From Stuart, B., *Biological Applications of Infrared spectroscopy*, ACOL Series, Wiley, Chichester, UK, 1997. © University of Greenwich, and reproduced by permission of the University of Greenwich

Component	Concentration (mg ml^{-1})	C=O stretching wavenumber (cm^{-1})	Absorbance
Aspirin	90	1764	0.217
Phenacetin	65	1511	0.185
Caffeine	15	1656	0.123

SAQ 3.5

Quantitative analysis of tablets containing aspirin, phenacetin and caffeine is to be carried out. Each of these components show distinct carbonyl bands in chloroform solution and calibration data for known concentrations are listed below in Table 3.4. Each of the standards was studied by using a 0.1 mm pathlength

NaCl transmission cell. Given that the absorbance values for an unknown tablet produced under the same conditions are given below in Table 3.5, estimate the concentrations of aspirin, phenacetin and caffeine in the unknown tablet.

Table 3.5 Absorbance values for an unknown drug mixture (cf. SAQ 3.5). From Stuart, B., *Biological Applications of Infrared spectroscopy*, ACOL Series, Wiley, Chichester, UK, 1997. © University of Greenwich, and reproduced by permission of the University of Greenwich

Wavenumber (cm^{-1})	Absorbance
1764	0.207
1511	0.202
1656	0.180

3.9 Calibration Methods

The improvement in computer technology associated with spectroscopy has led to the expansion of quantitative infrared spectroscopy. The application of statistical methods to the analysis of experimental data is known as *chemometrics* [5–9]. A detailed description of this subject is beyond the scope of this present text, although several multivariate data analytical methods which are used for the analysis of FTIR spectroscopic data will be outlined here, without detailing the mathematics associated with these methods. The most commonly used analytical methods in infrared spectroscopy are classical least-squares (CLS), inverse least-squares (ILS), partial least-squares (PLS), and principal component regression (PCR). CLS (also known as K-matrix methods) and PLS (also known as P-matrix methods) are least-squares methods involving matrix operations. These methods can be limited when very complex mixtures are investigated and factor analysis methods, such as PLS and PCR, can be more useful. The factor analysis methods use functions to model the variance in a data set.

If it is not necessary to know the specific concentrations of the species of interest, but to simply know whether such species are present, or not, in a complex sample; then, a multivariate pattern recognition method, such as linear discriminant analysis (LDA) or artificial neural networks (ANNs), may be used to identify the spectral characteristics of the species of interest [10]. Such methods are capable of comparing a large number of variables within a data set, such as intensity, frequency and bandwidth. LDA and ANNs are known as *supervised* methods because *a priori* information is available about the data sets. There are also *unsupervised* methods, such as hierarchical clustering, which can be used to determine the components in a data set without any prior information about the data.

The use of calibration methods in infrared spectroscopy is illustrated here by an example of the application of PCR to the analysis of vegetable oils [11]. Figure 3.15 shows the superposition of the infrared spectra of five different types of vegetable oil. This demonstrates the similarity of these spectra and the difficulty in differentiating the samples by simple inspection. However, there are characteristic differences in the band positions of each spectrum which may be exploited by using PCR; the wavenumbers of the C–H stretching and C–H bending bands and bands in the fingerprint region can be used as variables. The process of analysis involves a series of steps. Replicate spectra should first be obtained for the oils to be analysed and for the unknown samples. The mean of each wavenumber for each class of oil should be calculated. The absolute difference between the unknown wavenumber and the means for each class and each band are then determined. Then, the sum of the differences for each class across all bands employed is determined and the assignment of the unknown to the class with the smallest sum of differences can be made. The results of the analysis of a mixture of unknown vegetable oils, with seven peaks being used, are shown in Figure 3.16. Different symbols are used here to represent each type of

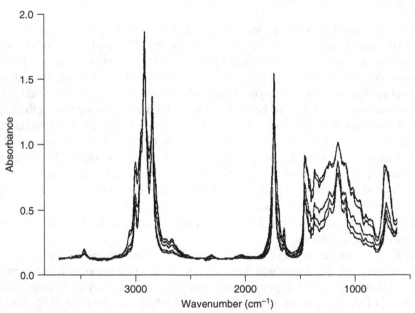

Figure 3.15 Infrared spectra of a mixture of five different vegetable oils [11]. Used with permission from the *Journal of Chemical Education*, **80**, No. 5, 2003, pp. 541–543; Copyright ©2003, Division of Chemical Education, Inc.

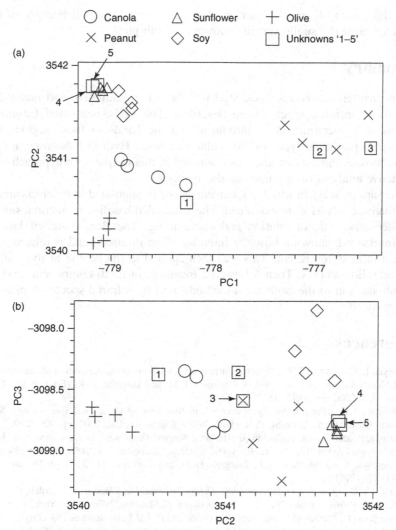

Figure 3.16 Plots of the principal component scores for vegetable oil analysis [11]. Used with permission from the *Journal of Chemical Education*, **80**, No. 5, 2003, pp. 541–543; Copyright ©2003, Division of Chemical Education, Inc.

known oil and each unknown oil. The plot illustrated in Figure 3.16(a) shows a PC2 versus PC1 plot and indicates that olive, canola and peanut oils form distinct clusters. This plot can be used to assign unknown '1' to the canola group and unknowns '2' and '3' to the peanut group. The plot illustrated in Figure 3.16(b) shows a PC3 versus PC2 plot. While sunflower and soy oil were not distinguishable in the PC2 versus PC1 plot, they are well separated in the PC3 versus PC2 plot. Unknowns '4' and '5' can be identified as sunflower oil because of their clustering near the sunflower oil 'knowns' in both plots.

Summary

In this chapter, the issues associated with both the qualitative and quantitative analysis of infrared spectra were described. The mid-, near- and far-infrared regions of a spectrum were introduced and the bands in these regions were assigned to particular types of molecular vibrations. Hydrogen bonding, a factor that influences infrared bands, was discussed in this chapter. An approach to the qualitative analysis of spectra was also described.

The various ways in which a spectrum can be manipulated in order to carry out quantitative analysis were examined. These included baseline correction, smoothing, derivatives, deconvolution and curve-fitting. The Beer–Lambert law was also introduced, showing how the intensity of an infrared band is related to the amount of analyte present. This was then applied to the simple analysis of liquid and solid samples. Then followed a treatment of multi-component mixtures. An introduction to the calibration methods used by infrared spectroscopists was also provided.

References

1. Weyer, L. G. and Lo, S. C., 'Spectra–Structure Correlations in the Near-infrared', in *Handbook of Vibrational Spectroscopy*, Vol. 3, Chalmers, J. M. and Griffiths, P. R. (Eds), Wiley, Chichester, UK, 2002, pp. 1817–1837.
2. Griffiths, P. R., 'Far-Infrared Spectroscopy', in *Handbook of Vibrational Spectroscopy*, Vol. 1, Chalmers, J. M. and Griffiths, P. R. (Eds), Wiley, Chichester, UK, 2002, pp. 229–239.
3. Debska, B. and Guzowska-Swide, E., 'Infrared Spectral Databases', in *Encyclopedia of Analytical Chemistry*, Vol. 12, Meyers, R. A. (Ed.), Wiley, Chichester, UK, 2000, pp. 10 928–10 953.
4. Kauppinen, J. K., Moffatt, D. J., Mantsch, H. H. and Cameron, D. G., *Appl. Spectrosc.*, **35**, 271–276 (1981).
5. Smith, B. C., *Quantitative Spectroscopy: Theory and Practice*, Elsevier, Amsterdam (2002).
6. Mark, H., *Principles and Practice of Spectroscopic Calibration*, Wiley, New York, 1996.
7. Hasegawa, T., 'Principal Component Regression and Partial Least Squares Modeling', in *Handbook of Vibrational Spectroscopy*, Vol. 3, Chalmers, J. M. and Griffiths, P. R. (Eds), Wiley, Chichester, UK, 2002, pp. 2293–2312.
8. Brown, S. D., 'Chemometrics', in *Encyclopedia of Analytical Chemistry*, Vol. 11, Meyers, R. A. (Ed.), Wiley, Chichester, UK, 2000, pp. 9671–9678.
9. Franke, J. E., 'Inverse Least Squares and Classical Least Squares Methods for Quantitative Vibrational Spectroscopy', in *Handbook of Vibrational Spectroscopy*, Vol. 3, Chalmers, J. M. and Griffiths, P. R. (Eds), Wiley, Chichester, UK, 2002, pp. 2276–2292.
10. Yang, H., 'Discriminant Analysis by Neural Networks', in *Handbook of Vibrational Spectroscopy*, Vol. 3, Chalmers, J. M. and Griffiths, P. R. (Eds), Wiley, Chichester, UK, 2002, pp. 2094–2106.
11. Rusak, D. A., Brown, L. M. and Martin, S. D., *J. Chem. Edu.*, **80**, 541–543 (2003).

Chapter 4
Organic Molecules

Learning Objectives

- To assign the infrared bands of the main classes of organic molecules.
- To identify the structures of organic molecules by using the infrared spectra of such molecules.

4.1 Introduction

One of the most common applications of infrared spectroscopy is to the identification of organic compounds. The major classes of organic molecules are examined in turn in this chapter and useful group frequencies are detailed. There are a number of useful collections of spectral information regarding organic molecules that may be used for reference and combined with the information provided in this chapter [1–7]. A series of questions are also provided to practice the assignment of bands observed in organic infrared spectra and to use the spectra to characterize the structures of organic molecules.

4.2 Aliphatic Hydrocarbons

Alkanes contain only C–H and C–C bonds, but there is plenty of information to be obtained from the infrared spectra of these molecules. The most useful are those arising from C–H stretching and C–H bending. C–H stretching bands in aliphatic

Infrared Spectroscopy: Fundamentals and Applications   B. Stuart
© 2004 John Wiley & Sons, Ltd   ISBNs: 0-470-85427-8 (HB); 0-470-85428-6 (PB)

hydrocarbons appear in the 3000–2800 cm^{-1} range and the C–H stretching bands of methyl groups and methylene groups are readily differentiated. For methyl groups, asymmetric C–H stretching occurs at 2870 cm^{-1}, while symmetric C–H stretching occurs at 2960 cm^{-1}. By comparison, methylene groups show asymmetric stretching at 2930 cm^{-1} and symmetric stretching at 2850 cm^{-1}. C–H bending gives rise to bands in the region below 1500 cm^{-1}. Methyl groups produce two bending bands, i.e. a symmetrical band at 1380 cm^{-1} and an asymmetrical band at 1475 cm^{-1}. Methylene groups give rise to four bending vibrations: scissoring (1465 cm^{-1}), rocking (720 cm^{-1}), wagging (1305 cm^{-1}) and twisting (1300 cm^{-1}). The intensity of the methylene CH$_2$ rocking band is useful as four or more CH$_2$ groups are required in a chain to produce a distinct band near 720 cm^{-1}. Shorter chains show a more variable band, for instance, the CH$_2$ rocking band for C$_4$H$_{10}$ is near 734 cm^{-1}. Although these are the main characteristic bands associated with aliphatic hydrocarbons, there are a number of bands that appear in the spectra of such compounds as there is a wide range of structures possible. The main infrared bands for alkanes are summarized in Table 4.1.

Table 4.1 Characteristic infrared bands of aliphatic hydrocarbons

Wavenumber (cm^{-1})	Assignment
	Alkanes
2960	Methyl symmetric C–H stretching
2930	Methylene asymmetric C–H stretching
2870	Methyl asymmetric C–H stretching
2850	Methylene symmetric C–H stretching
1470	Methyl asymmetrical C–H bending
1465	Methylene scissoring
1380	Methyl symmetrical C–H bending
1305	Methylene wagging
1300	Methylene twisting
720	Methylene rocking
	Alkenes
3100–3000	=C–H stretching
1680–1600	C=C stretching
1400	=C–H in-plane bending
1000–600	=C–H out-of-plane bending
	Alkynes
3300–3250	≡C–H stretching
2260–2100	C≡C stretching
700–600	≡C–H bending

SAQ 4.1

Examine the infrared spectrum of nonane shown below in Figure 4.1 and describe the vibrations corresponding to the bands marked A, B and C.

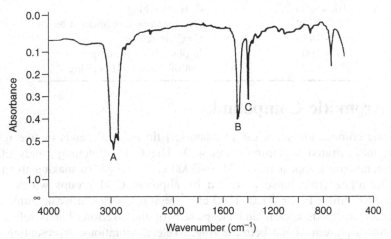

Figure 4.1 Infrared spectrum of nonane (C_9H_{20}) (cf. SAQ 4.1). From Stuart, B., *Modern Infrared Spectroscopy*, ACOL Series, Wiley, Chichester, UK, 1996. © University of Greenwich, and reproduced by permission of the University of Greenwich.

SAQ 4.2

From the following list of compounds, 2-methylbutane, cyclopentane, 2-methyloctane, 3-methylpentane, butane and 1,1-dimethylcyclohexane, choose those that:

(a) absorb at 720 cm^{-1};
(b) absorb at 1375 cm^{-1};
(c) absorb at both 720 and 1375 cm^{-1};
(d) do not absorb at either 720 or 1375 cm^{-1}.

Alkenes contain the C=C group, with the majority having hydrogen attached to the double bond. Four major bands are associated with this molecular fragment. These are the out-of-plane and in-plane =C–H deformations, C=C stretching and =C–H stretching. The characteristic infrared modes observed for alkenes are given in Table 4.1.

Alkynes contain the C≡C group and three characteristic bands can be present, i.e. ≡C–H stretching, ≡C–H bending and C≡C stretching. The main bands observed for alkynes are also shown in Table 4.1.

Table 4.2 Characteristic infrared bands of aromatic compounds. From Stuart, B., *Modern Infrared Spectroscopy*, ACOL Series, Wiley, Chichester, UK, 1996. © University of Greenwich, and reproduced by permission of the University of Greenwich

Wavenumber (cm^{-1})	Assignment
3100–3000	C–H stretching
2000–1700	Overtone and combination bands
1600–1430	C=C stretching
1275–1000	In-plane C–H bending
900–690	Out-of-plane C–H bending

4.3 Aromatic Compounds

Aromatic compounds show useful characteristic infrared bands in five regions of the mid-infrared spectrum (Table 4.2). The C–H stretching bands of aromatic compounds appear in the 3100–3000 cm^{-1} range, so making them easy to differentiate from those produced by aliphatic C–H groups which appear below 3000 cm^{-1}. In the 2000–1700 cm^{-1} region, a series of weak combination and overtone bands appear and the pattern of the overtone bands reflects the substitution pattern of the benzene ring. Skeletal vibrations, representing C=C stretching, absorb in the 1650–1430 cm^{-1} range. The C–H bending bands appear in the regions 1275–1000 cm^{-1} (in-plane bending) and 900–690 cm^{-1} (out-of-plane bending). The bands of the out-of-plane bending vibrations of aromatic compounds are strong and characteristic of the number of hydrogens in the

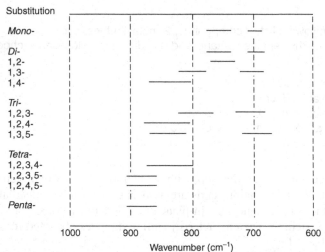

Figure 4.2 Out-of-plane bending vibrations of aromatic compounds. From Stuart, B., *Modern Infrared Spectroscopy*, ACOL Series, Wiley, Chichester, UK, 1996. © University of Greenwich, and reproduced by permission of the University of Greenwich.

ring, and hence can be used to give the substitution pattern. This information is summarized in Figure 4.2.

SAQ 4.3

Examine the IR spectra of three isomeric disubstituted benzenes presented below in Figures 4.3, 4.4 and 4.5. Which of these is 1,2-, which 1,3- and which 1,4-disubstituted?

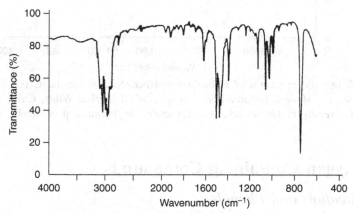

Figure 4.3 Infrared spectrum of a disubstituted benzene: compound A (cf. SAQ 4.3). From Stuart, B., *Modern Infrared Spectroscopy*, ACOL Series, Wiley, Chichester, UK, 1996. © University of Greenwich, and reproduced by permission of the University of Greenwich.

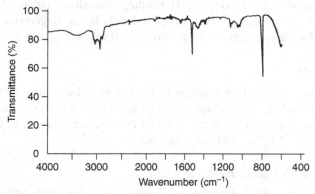

Figure 4.4 Infrared spectrum of a disubstituted benzene: compound B (cf. SAQ 4.3). From Stuart, B., *Modern Infrared Spectroscopy*, ACOL Series, Wiley, Chichester, UK, 1996. © University of Greenwich, and reproduced by permission of the University of Greenwich.

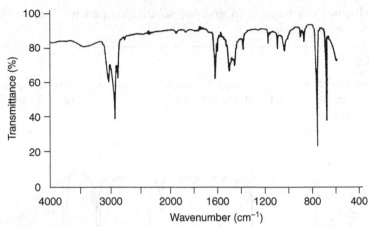

Figure 4.5 Infrared spectrum of a disubstituted benzene: compound C (cf. SAQ 4.3). From Stuart, B., *Modern Infrared Spectroscopy*, ACOL Series, Wiley, Chichester, UK, 1996. © University of Greenwich, and reproduced by permission of the University of Greenwich.

4.4 Oxygen-Containing Compounds

4.4.1 Alcohols and Phenols

Alcohols and phenols produce characteristic infrared bands due to O–H stretching and C–O stretching, which are both sensitive to hydrogen bonding. For alcohols, the broad O–H stretching band is centred at 3600 cm^{-1}, while for phenols this band appears 50–100 cm^{-1} lower than the alcohol wavenumber. C–O stretching in alcohols and phenols produces a strong band in the 1300–1000 cm^{-1} region. These compounds also produce O–H bending vibrations, but the latter couple with other vibrations and produce complex bands in the fingerprint region. The main bands for these compounds are shown in Table 4.3.

4.4.2 Ethers

Ethers may be identified by a strong C–O stretching band near 1100 cm^{-1} due to the C–O–C linkage in this type of compound (see Table 4.3). Aromatic ethers show a strong band near 1250 cm^{-1}, while cyclic ethers show a C–O stretching band over a broad 1250–900 cm^{-1} range.

4.4.3 Aldehydes and Ketones

Aliphatic and aromatic ketones show carbonyl bands at 1730–1700 and 1700–1680 cm^{-1}, respectively, while aliphatic and aromatic aldehydes produce carbonyl bands in the 1740–1720 cm^{-1} and 1720–1680 cm^{-1} ranges, respectively (see Table 4.3). The position of the C=O stretching wavenumber within these

Table 4.3 Characteristic infrared bands of oxygen-containing compounds

Wavenumber (cm^{-1})	Assignment
	Alcohol and phenols
3600	Alcohol O–H stretching
3550–3500	Phenol O–H stretching
1300–1000	C–O stretching
	Ethers
1100	C–O–C stretching
	Aldehydes and ketones
2900–2700	Aldehyde C–H stretching
1740–1720	Aliphatic aldehyde C=O stretching
1730–1700	Aliphatic ketone C=O stretching
1720–1680	Aromatic aldehyde C=O stretching
1700–1680	Aromatic ketone C=O stretching
	Esters
1750–1730	Aliphatic C=O stretching
1730–1705	Aromatic C=O stretching
1310–1250	Aromatic C–O stretching
1300–1100	Aliphatic C–O stretching
	Carboxylic acids
3300–2500	O–H stretching
1700	C=O stretching
1430	C–O–H in-plane bending
1240	C–O stretching
930	C–O–H out-of-plane bending
	Anhydrides
1840–1800	C=O stretching
1780–1740	C=O stretching
1300–1100	C–O stretching

ranges is dependent on hydrogen bonding and conjugation within the molecule. Conjugation with a C=C band results in delocalization of the C=O group, hence causing the absorption to shift to a lower wavenumber. Aldehydes also show a characteristic C–H stretching band in the 2900–2700 cm^{-1} range.

DQ 4.1

Examine the spectrum of 2-methylbenzaldehyde shown below in Figure 4.6 and identify the aldehyde C–H stretching band. Describe the effect responsible for the doublet formation.

Answer

*Aldehydes show C–H stretching in the 2900–2700 cm^{-1} region. **Fermi
resonance** is responsible for the formation of the doublet. Figure 4.6
shows a doublet centred at 2780 cm^{-1} which may be assigned to aldehyde
C–H stretching. This is the coupling of a **fundamental** with an **overtone**
band, with the two vibrations being in the same place. As the doublet
is centred at about 2780 cm^{-1}, a band near 1400 cm^{-1} is expected to
be the fundamental. There is a doublet in this region with peaks at 1382
and 1386 cm^{-1}. One of these is a CH$_3$ bending band, while the other is
an H–C–O in-plane bending band. The latter must be responsible for the
doublet since the CH$_3$ bending band is not in the same plane as the aldehyde C–H stretching movement. Twice 1386 cm^{-1} is 2772 cm^{-1}, which
is close to the centre of the observed doublet.*

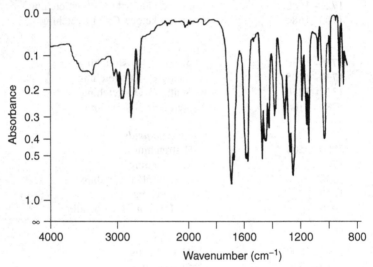

Figure 4.6 Infrared spectrum of 2-methoxybenzaldehyde (cf. DQ 4.1).

4.4.4 Esters

The two most polar bonds in esters (containing the –CO–O–C–unit) are the C=O
and C–O bonds and these bonds produce the strongest bands in the spectrum of
any ester (summarized in Table 4.3). Aromatic and aliphatic esters may be differentiated, as both C=O stretching and C–O stretching vibrations produce bands
in different ranges: aliphatic esters produce C=O and C–O bands at 1750–1730
and 1300–1100 cm^{-1}, respectively, while aromatic esters produce C=O and C–O
bands at 1730–1705 and 1310–1250 cm^{-1}, respectively.

4.4.5 Carboxylic Acids and Anhydrides

Carboxylic acids (RCOOH) exist as dimers, except in dilute solution, due to strong intermolecular hydrogen bonding. Carboxylic acids show a strong broad O–H stretching band in the 3300–2500 cm^{-1} range. The C=O stretching band of the dimer is observed near 1700 cm^{-1}, while the free acid band is observed at higher wavenumbers (1760 cm^{-1}). In addition, carboxylic acids show characteristic C–O stretching and in-plane and out-of-plane O–H bending bands at 1240, 1430 and 930 cm^{-1}, respectively.

Anhydrides (–CO–O–CO–) may be identified by a distinctive carbonyl band region. A strong doublet is observed in this region with components in the 1840–1800 and 1780–1740 cm^{-1} ranges. Anhydrides also show a strong C–O stretching band near 1150 cm^{-1} for open-chain anhydrides and at higher wavenumbers for cyclic structures. Table 4.3 summarizes the major infrared bands observed for carboxylic acids and anhydrides.

SAQ 4.4

Assign the following pairs of carbonyl stretching wavenumbers to the corresponding structures shown below in Figure 4.7: 1815 and 1750 cm^{-1}, and 1865 and 1780 cm^{-1}.

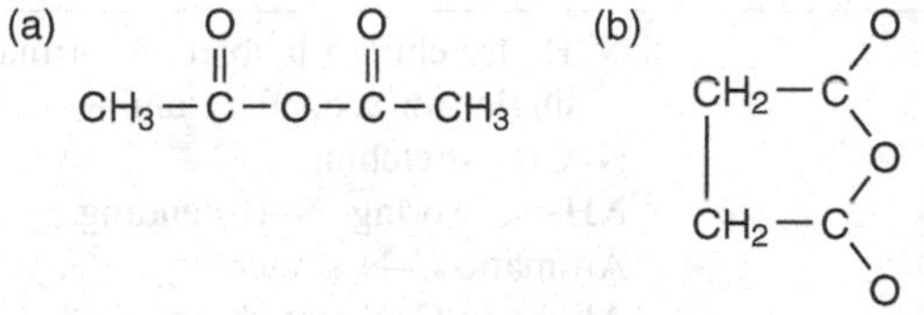

(a)

$$CH_3-\overset{\overset{O}{\|}}{C}-O-\overset{\overset{O}{\|}}{C}-CH_3$$

(b)

Figure 4.7 Anhydride structures (cf. SAQ 4.4).

DQ 4.2

Given that a molecule contains oxygen, how could an infrared spectrum of such a molecule be used to find out whether the compound was a carboxylic acid, a phenol, an alcohol or an ether?

Answer

Ethers can easily be identified because they are the only compounds of the four that do not absorb above 3200 cm^{-1}. Carboxylic acids have a very broad absorption from the O–H stretching between 2500 and 3500 cm^{-1}. They are readily distinguished from alcohols and phenols where the O–H absorption bands are not as broad as those of carboxylic acids. Both *phenols and alcohols absorb in the O–H stretching region and can only be differentiated by the benzene ring absorption of the phenol. If the alcohol is an aromatic alcohol, a new approach is required and an additional chemical test will be required.*

4.5 Nitrogen-Containing Compounds

4.5.1 Amines

Primary ($-NH_2$), secondary ($-NH$) and tertiary (no hydrogen attached to N) amines may be differentiated by using infrared spectra. Primary amines have two sharp N–H stretching bands near 3335 cm^{-1}, a broad NH_2 scissoring band at 1615 cm^{-1}, and NH_2 wagging and twisting bands in the 850–750 cm^{-1} range. Secondary amines show only one N–H stretching band at 3335 cm^{-1} and an N–H bending band at 1615 cm^{-1}. The N–H wagging band for secondary amines appears at 715 cm^{-1}. Tertiary amines are characterized by an N–CH_2 stretching band at 2780 cm^{-1}. All amines show C–N stretching bands, with aromatic amines showing such bands in the 1360–1250 cm^{-1} range and aliphatic amines showing bands at 1220–1020 cm^{-1}. Table 4.4 summarizes the main infrared bands for amines.

Table 4.4 Characteristic infrared bands of amines

Wavenumber (cm^{-1})	Assignment
3335	N–H stretching (doublet for primary amines; singlet for secondary amines)
2780	N–CH_2 stretching
1615	NH_2 scissoring, N–H bending
1360–1250	Aromatic C–N stretching
1220–1020	Aliphatic C–N stretching
850–750	NH_2 wagging and twisting
715	N–H wagging

SAQ 4.5

Figure 4.8 below shows the infrared spectrum of a liquid film of hexylamine. Assign the major bands appearing in this spectrum.

4.5.2 Amides

Primary amides ($-CO-NH-$) display two strong NH_2 stretching bands, i.e. asymmetric stretching at 3360–3340 cm^{-1} and symmetric stretching at

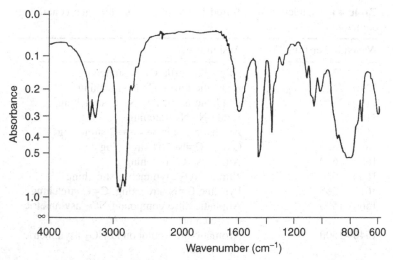

Figure 4.8 Infrared spectrum of hexylamine (cf. SAQ 4.5). From Stuart, B., *Modern Infrared Spectroscopy*, ACOL Series, Wiley, Chichester, UK, 1996. © University of Greenwich, and reproduced by permission of the University of Greenwich.

$3190–3170$ cm^{-1}. They also exhibit C=O stretching at $1680–1660$ cm^{-1} (referred to as the amide I band) and NH$_2$ bending at $1650–1620$ cm^{-1} (referred to as the amide II band). Secondary amides show an N–H stretching band at $3300–3250$ cm^{-1}, while the carbonyl stretching (amide I band) is observed at $1680–1640$ cm^{-1}. The amide II band for secondary amides is due to the coupling of N–H bending and C–N stretching and appears at $1560–1530$ cm^{-1}. A weak band which is an overtone of the amide II band appears at $3100–3060$ cm^{-1}. A broad N–H wagging band also appears at $750–650$ cm^{-1}. Table 4.5 summarizes the infrared bands of primary and secondary amides.

Table 4.5 Characteristic infrared bands of amides

Wavenumber (cm^{-1})	Assignment
3360–3340	Primary amide NH$_2$ asymmetric stretching
3300–3250	Secondary amide N–H stretching
3190–3170	Primary amide NH$_2$ symmetric stretching
3100–3060	Secondary amide amide II overtone
1680–1660	Primary amide C=O stretching
1680–1640	Secondary amide C=O stretching
1650–1620	Primary amide NH$_2$ bending
1560–1530	Secondary amide N–H bending, C–N stretching
750–650	Secondary amide N–H wagging

Table 4.6 Characteristic infrared bands of various nitrogen-containing compounds

Wavenumber (cm^{-1})	Assignment
2260–2240	Aliphatic nitrile C≡N stretching
2240–2220	Aromatic nitrile C≡N stretching
2180–2110	Aliphatic isonitrile –N≡C stretching
2160–2120	Azide N≡N stretching
2130–2100	Aromatic isonitrile –N≡C stretching
1690–1620	Oxime C=N–OH stretching
1680–1650	Nitrite N=O stretching
1660–1620	Nitrate NO$_2$ asymmetric stretching
1615–1565	Pyridine C=N stretching, C=C stretching
1560–1530	Aliphatic nitro compound NO$_2$ asymmetric stretching
1540–1500	Aromatic nitro compound NO$_2$ asymmetric stretching
1450–1400	Azo compound N=N stretching
1390–1370	Aliphatic nitro compound NO$_2$ symmetric stretching
1370–1330	Aromatic nitro compound NO$_2$ symmetric stretching
1300–1270	Nitrate NO$_2$ symmetric stretching
965–930	Oxime N–O stretching
870–840	Nitrate N–O stretching
710–690	Nitrate NO$_2$ bending

Other nitrogen-containing compounds, such as nitriles, azides, oximes, nitrites and pyridines, also show characteristic infrared bands which may be used for identification purposes. These bands are summarized in Table 4.6.

4.6 Halogen-Containing Compounds

Table 4.7 lists the absorption regions observed for organic halogen compounds. Particular care must be taken when fluorine is present in a compound as the C–F stretching bands, observed between 1350 and 1100 cm^{-1}, are very strong and may obscure any C–H bands that might be present.

Table 4.7 Characteristic infrared bands of organic halogen compounds

Wavenumber (cm^{-1})	Assignment
1300–1000	C–F stretching
800–400	C–X stretching (X = F, Cl, Br or I)

4.7 Heterocyclic Compounds

Heterocyclic compounds, such as pyridines, pyrazines, pyrroles and furanes, show C–H stretching bands in the 3080–3000 cm^{-1} region. For heterocyclic compounds containing an N–H group, an N–H stretching band is observed in the 3500–3200 cm^{-1} region, the position of which depends upon the degree of hydrogen bonding. These compounds show characteristic band patterns in the 1600–1300 cm^{-1} ring stretching region. The nature of the substituents affects the pattern. In addition, the pattern of out-of-plane C–H bending bands in the 800–600 cm^{-1} region are characteristic for heterocyclic molecules.

4.8 Boron Compounds

Compounds containing the B–O linkage, such as boronates and boronic acids, are characterized by a strong B–O stretching mode at 1380–1310 cm^{-1}. Boronic acid and boric acid contain OH groups and so show a broad O–H stretching mode in the 3300–3200 cm^{-1} region. Compounds with a B–N group, such as borazines and amino boranes, show a strong absorption in the 1465–1330 cm^{-1} region due to B–N stretching. B–H stretching, due to BH and BH$_2$ groups, produces bands in the 2650–2350 cm^{-1} range. The BH$_2$ absorption is usually a doublet due to symmetric and asymmetric vibrations. There is also a B–H bending band in the 1205–1140 cm^{-1} region and a wagging band at 980–920 cm^{-1} due to the BH$_2$ group. Table 4.8 provides a summary of the main infrared modes in boron compounds.

4.9 Silicon Compounds

Silicon compounds show characteristic Si–H bands. The Si–H stretching bands appear at 2250–2100 cm^{-1}. The Si–CH$_3$ band gives rise to a symmetric CH$_3$

Table 4.8 Characteristic infrared bands of boron compounds. From Stuart, B., *Modern Infrared Spectroscopy*, ACOL Series, Wiley, Chichester, UK, 1996. © University of Greenwich, and reproduced by permission of the University of Greenwich

Wavenumber (cm^{-1})	Assignment
3300–3200	B–O–H stretching
2650–2350	B–H stretching
1465–1330	B–N stretching
1380–1310	B–O stretching
1205–1140	B–H bending
980–920	B–H wagging

Table 4.9 Characteristic infrared bands of silicon compounds

Wavenumber (cm^{-1})	Assignment
3700–3200	Si–OH stretching
2250–2100	Si–H stretching
1280–1250	Si–CH$_3$ symmetric bending
1430, 1110	Si–C$_6$H$_5$ stretching
1130–1000	Si–O–Si asymmetric stretching
1110–1050	Si–O–C stretching

bending band in the 1280–1250 cm^{-1} range. The O–H stretching bands of Si–OH groups appear in the same region as alcohols, i.e. 3700–3200 cm^{-1}. Si–O–C stretching produces a broad band at 1100–1050 cm^{-1} and siloxanes also show a strong band at 1130–1000 cm^{-1}. Silicon attached to a benzene ring produces two strong bands near 1430 and 1110 cm^{-1}. Some of the common infrared modes of silicon compounds are summarized in Table 4.9.

4.10 Phosphorus Compounds

Phosphorus acids and esters possess P–OH groups which produce one or two broad bands in the 2700–2100 cm^{-1} region. Aliphatic phosphorus compounds

Table 4.10 Characteristic infrared bands of phosphorus compounds

Wavenumber (cm^{-1})	Assignment
2425–2325	Phosphorus acid and ester P–H stretching
2320–2270	Phosphine P–H stretching
1090–1080	Phosphine PH$_2$ bending
990–910	Phosphine P–H wagging
2700–2100	Phosphorus acid and ester O–H stretching
1040–930	Phosphorus ester P–OH stretching
1050–950	Aliphatic asymmetric P–O–C stretching
830–750	Aliphatic symmetric P–O–C stretching
1250–1160	Aromatic P–O stretching
1050–870	Aromatic P–O stretching
1450–1430	Aromatic P–C stretching
1260–1240	Aliphatic P=O stretching
1350–1300	Aromatic P=O stretching
1050–700	P–F stretching
850–500	P=S stretching
600–300	P–Cl stretching
500–200	P–Br stretching
500–200	P–S stretching

containing a P–O–C link show a strong band between 1050 and 950 cm^{-1} due to asymmetric stretching. If a P=O group is present in an aliphatic phosphorus compound, a strong band near 1250 cm^{-1} is observed. Aromatic phosphorus compounds show characteristic bands that enable them to be differentiated from aliphatic compounds. A strong sharp band near 1440 cm^{-1} appears for compounds where the phosphorus atom is directly attached to the benzene ring. Where the P–O group is attached to the ring, two bands appear in the 1250–1160 cm^{-1} and 1050–870 cm^{-1} regions due to P–O stretching. When the P–O group is attached to the ring via the oxygen, two bands also appear, but the higher-wavenumber band is observed between 1350 and 1250 cm^{-1}. Table 4.10 summarizes the important infrared bands of some common phosphorus compounds.

SAQ 4.6

Figure 4.9 below shows the spectrum of a phosphorus compound with the empirical formula $C_2H_7PO_3$. Assign the major infrared bands in this spectrum.

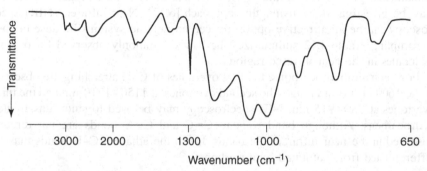

Figure 4.9 Infrared spectrum of an unknown phosphorus compound ($C_2H_7PO_3$) (cf. SAQ 4.6). Reproduced by permission of J. Emsley and D. Hall, from *The Chemistry of Phosphorus*, Harper and Row, London, p. 107 (1976).

4.11 Sulfur Compounds

The SO_2 and SO groups of sulfur compounds produce strong infrared bands in the 1400–1000 cm^{-1} range. The expected S=O stretching bands for sulfoxides, sulfones, sulfonic acids, sulfonamides, sulfonyl chlorides and sulfonates are given in Table 4.11. The other bonds involving sulfur, such as S–H, S–S and C–S, produce weaker infrared bands. Compounds, such as sulfides and mercaptans,

Table 4.11 Characteristic infrared bands of sulfur compounds

Wavenumber (cm^{-1})	Assignment
700–600	C–S stretching
550–450	S–S stretching
2500	S–H stretching
1390–1290	SO$_2$ asymmetric stretching
1190–1120	SO$_2$ symmetric stretching
1060–1020	S=O stretching

containing C–S and S–S bonds, show stretching bands at 700–600 cm^{-1} and near 500 cm^{-1}, respectively. The weak S–H stretching band appears near 2500 cm^{-1}.

4.12 Near-Infrared Spectra

As near-infrared spectra are, for the most part, the result of overtone bands of fundamental groups containing C–H, O–H and N–H bonds, organic molecules may be investigated by using this approach [6, 7]. Near-infrared (NIR) spectroscopy can be an attractive option for certain organic systems because of ease of sampling. Table 4.12 summarizes the bands commonly observed for organic molecules in the near-infrared region.

For aliphatic hydrocarbons, the first overtones of C–H stretching are observed in the 1600–1800 nm range, the second overtones at 1150–1210 nm and the third overtones at 880–915 nm. NIR spectroscopy may be used to study unsaturated hydrocarbons. Although bands due to C=C and C≡C bonds are not directly observed in the near infrared, the bands due to the adjacent C–H bonds may be differentiated from saturated hydrocarbons.

Table 4.12 Common near-infrared bands of organic compounds

Wavelength (nm)	Assignment
2200–2450	Combination C–H stretching
2000–2200	Combination N–H stretching, combination O–H stretching
1650–1800	First overtone C–H stretching
1400–1500	First overtone N–H stretching, first overtone O–H stretching
1300–1420	Combination C–H stretching
1100–1225	Second overtone C–H stretching
950–1100	Second overtone N–H stretching, second overtone O–H stretching
850–950	Third overtone C–H stretching
775–850	Third overtone N–H stretching

Aromatic hydrocarbons produce different bands in the near infrared to those of aliphatic hydrocarbons. The first and second overtone C–H stretching bands for aromatic compounds are seen in the 1600–1800 nm and 1100–1250 nm regions, respectively. They also show a series of combination bands at 2100–2250 nm and 2450–2500 nm. Ring-substitution patterns affect the positions of these bands.

SAQ 4.7

Figure 4.10 below shows the NIR spectrum of benzene. Identify the major bands that appear in this spectrum.

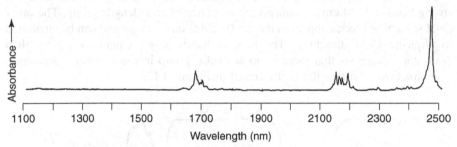

Figure 4.10 NIR spectrum of benzene (cf. SAQ 4.7). From Weyer, L. G. and Lo, S. C., 'Spectra–Structure Correlations in the Near-Infrared', in *Handbook of Vibrational Spectroscopy*, Vol. 3, Chalmers, J. M. and Griffiths, P. R. (Eds), pp. 1817–1837. Copyright 2002. © John Wiley & Sons Limited. Reproduced with permission.

For organic compounds containing C–O bonds, these groups affect the spectra by causing shifts in the bands of adjacent C–H, O–H and N–H groups. Alcohols and organic acids show combination and overtone O–H stretching bands in the 1900–2200 nm and 1400–1650 nm regions. The positions and shapes of these bands are significantly influenced by the degree of hydrogen bonding in such molecules.

Primary amines show characteristic first overtone bands at 1450, 1550 and 1000 nm which represent the first and second overtones of N–H stretching bands. Tertiary amines produce no N–H overtones, but do produce C–H and O–H stretching combination bands at 1260–1270 nm. The first overtones of the N–H stretching bands of aliphatic amines are shifted to lower wavelengths. Primary amides may be characterized by combination bands at 1930–2250 nm and an N–H stretching overtone band at 1450–1550 nm. Secondary amides produce overtone bands in the 1350–1550 nm region and combination bands in the 1990–2250 nm region.

4.13 Identification

This section provides a number of examples to illustrate approaches to the identification of unknown organic materials by using infrared spectroscopy. First, an example of the use of the technique in identifying an organic compound containing a number of functional groups is provided.

Citronellal is the terpenoid responsible for the characteristic aroma of lemon oil, and is used in perfumes and as a mosquito repellent. The infrared spectrum of citronellal is shown in Figure 4.11. At the higher-wavenumber end of the spectrum, the C–H stretching region provides important information about the citronellal structure. This region is intense, hence indicating the presence of a number of C–H groups in the structure. Distinct C–H stretching modes at 2800 and 2700 cm^{-1} appear at wavenumbers characteristic of aldehydes. In addition, a strong band at 1730 cm^{-1} confirms the presence of an aldehyde group. The other C–H stretching modes appear in the 3000–2900 cm^{-1} range and can be attributed to aliphatic C–H stretching. The lack of bands at wavenumbers greater than 3000 cm^{-1} suggests that there is no aromatic group in the citronellal structure. The structure of citronellal is illustrated in Figure 4.12.

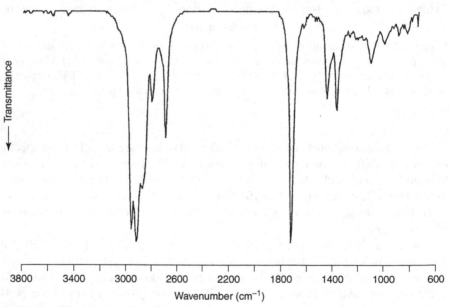

Figure 4.11 Infrared spectrum of citronellal. From Stuart, B., *Biological Applications of Infrared Spectroscopy*, ACOL Series, Wiley, Chichester, UK, 1997. © University of Greenwich, and reproduced by permission of the University of Greenwich.

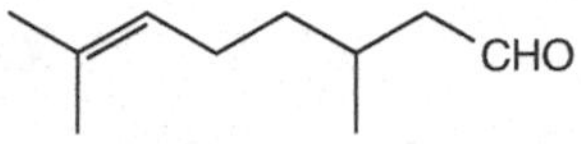

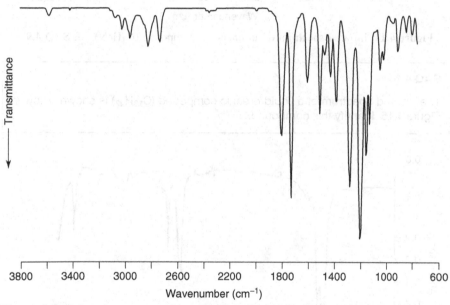

Figure 4.12 Structure of citronellal.

A series of SAQs is now provided for practice in the assignment of infrared bands and for the determination of structures by using infrared spectra.

SAQ 4.8

The infrared spectrum of vanillin is shown below in Figure 4.13. Vanillin occurs naturally in vanilla and is used as a flavouring agent. Identify the functional groups in this molecule.

Figure 4.13 Infrared spectrum of vanillin (cf. SAQ 4.8). From Stuart, B., *Biological Applications of Infrared Spectroscopy*, ACOL Series, Wiley, Chichester, UK, 1997. © University of Greenwich, and reproduced by permission of the University of Greenwich.

SAQ 4.9

Figure 4.14 below illustrates the infrared spectrum of a liquid film of a compound with the structural formula C_7H_9N. Assign the major infrared bands of this compound, and suggest a possible structure consistent with this spectrum.

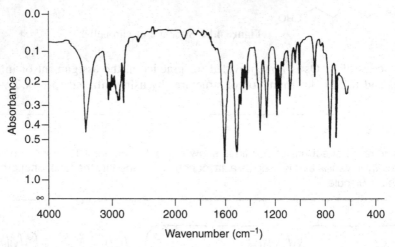

Figure 4.14 Infrared spectrum of an unknown compound (C_7H_9N) (cf. SAQ 4.9).

SAQ 4.10

The infrared spectrum of a liquid organic compound ($C_{10}H_{22}$) is shown below in Figure 4.15. Identify this compound.

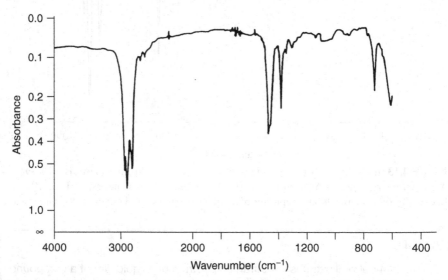

Figure 4.15 Infrared spectrum of an unknown hydrocarbon ($C_{10}H_{22}$) (cf. SAQ 4.10). From Stuart, B., *Modern Infrared Spectroscopy*, ACOL Series, Wiley, Chichester, UK, 1996. © University of Greenwich, and reproduced by permission of the University of Greenwich.

SAQ 4.11

The infrared spectrum of an organic compound ($C_4H_{10}O$) is shown below in Figure 4.16. Identify this compound.

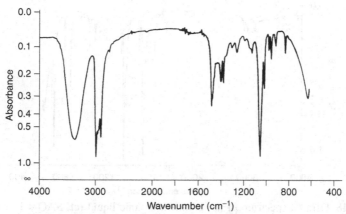

Figure 4.16 Infrared spectrum of an unknown liquid ($C_4H_{10}O$) (cf. SAQ 4.11). From Stuart, B., *Modern Infrared Spectroscopy*, ACOL Series, Wiley, Chichester, UK, 1996. © University of Greenwich, and reproduced by permission of the University of Greenwich.

SAQ 4.12

The infrared spectrum of an organic compound (C_8H_{16}) is shown below in Figure 4.17. Identify this compound.

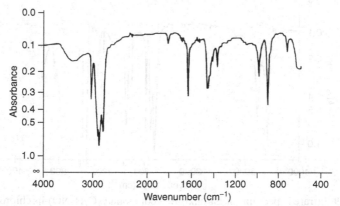

Figure 4.17 Infrared spectrum of an unknown hydrocarbon (C_8H_{16}) (cf. SAQ 4.12). From Stuart, B., *Modern Infrared Spectroscopy*, ACOL Series, Wiley, Chichester, UK, 1996. © University of Greenwich, and reproduced by permission of the University of Greenwich.

SAQ 4.13

The infrared spectrum of an organic compound is shown below in Figure 4.18. Identify this compound.

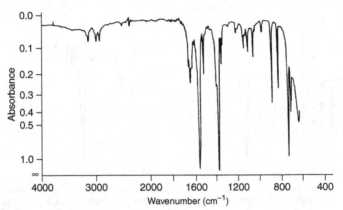

Figure 4.18 Infrared spectrum of an unknown organic liquid (cf. SAQ 4.13). From Stuart, B., *Modern Infrared Spectroscopy*, ACOL Series, Wiley, Chichester, UK, 1996. © University of Greenwich, and reproduced by permission of the University of Greenwich.

SAQ 4.14

The infrared spectrum of an organic compound (C_7H_7NO) in chloroform solution is shown below in Figure 4.19. Identify this compound.

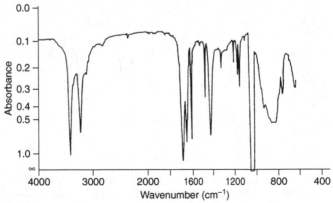

Figure 4.19 Infrared spectrum of an unknown compound (C_7H_7NO) in chloroform solution (cf. SAQ 4.14). From Stuart, B., *Modern Infrared Spectroscopy*, ACOL Series, Wiley, Chichester, UK, 1996. © University of Greenwich, and reproduced by permission of the University of Greenwich.

Summary

This chapter has provided a review of the major infrared bands associated with common classes of organic molecules. Examples were provided in order to use this information to assign the infrared bands of a range of organic molecules and to determine the structures of such molecules. This information may then be further utilized to characterize the infrared spectra obtained for various other organic compounds, for example, polymeric and biological materials and those used in industrial applications. The latter are introduced in Chapters 6, 7 and 8, respectively.

References

1. Silverstein, R. M. and Webster, F. X., *Spectrometric Identification of Organic Compounds*, 6th Edn, Wiley, New York, 1997.
2. Günzler, H. and Gremlich, H.-U., *IR Spectroscopy: An Introduction*, Wiley-VCH, Weinheim, Germany, 2002.
3. Linvien, D., *The Handbook of Infrared and Raman Characteristic Frequencies of Organic Molecules*, Academic Press, Boston, MA, USA, 1991.
4. Shurvell, H. F., 'Spectra–Structure Correlations in the Mid- and Far-Infrared', in *Handbook of Vibrational Spectroscopy*, Vol. 3, Chalmers, J. M. and Griffiths, P. R. (Eds), Wiley, Chichester, UK, 2002, pp. 1783–1816.
5. Coates, J., 'Interpretation of Infrared Spectra, A Practical Approach', in *Encyclopedia of Analytical Chemistry*, Vol. 12, Meyers, R. A. (Ed.), Wiley, Chichester, UK, 2000, pp. 10815–10837.
6. Weyer, L. G. and Lo, S. C., 'Spectra–Structure Correlations in the Near-infrared', in *Handbook of Vibrational Spectroscopy*, Vol. 3, Chalmers, J. M. and Griffiths, P. R. (Eds), Wiley, Chichester, UK, 2002, pp. 1817–1837.
7. Miller, C. E., 'Near-infrared Spectroscopy of Synthetic and Industrial Samples', in *Handbook of Vibrational Spectroscopy*, Vol. 1, Chalmers, J. M. and Griffiths, P. R. (Eds), Wiley, Chichester, UK, 2002, pp. 196–211.

Chapter 5

Inorganic Molecules

Learning Objectives

- To recognize the characteristic infrared bands of inorganic ions.
- To appreciate the effect of hydration on the infrared spectra of inorganic molecules.
- To understand the normal modes of vibration of inorganic molecules of different structural types.
- To understand the effects of coordination of inorganic ions on the infrared spectra of such compounds.
- To understand the effects of isomerism on the infrared spectra of coordination compounds.
- To understand how the bonding in metal carbonyl compounds affects the infrared spectra of such compounds.
- To appreciate the infrared bands associated with organometallic compounds.
- To use the infrared spectra of minerals to understand the structural properties of such materials.

5.1 Introduction

As with organic compounds, inorganic compounds can produce an infrared spectrum. Generally, the infrared bands for inorganic materials are broader, fewer in number and appear at lower wavenumbers than those observed for organic materials. If an inorganic compound forms covalent bonds within an ion, it can produce a characteristic infrared spectrum. The bands in the spectrum of ionic or coordination compounds will depend on the structure and orientation of the ion

---

Infrared Spectroscopy: Fundamentals and Applications   B. Stuart
© 2004 John Wiley & Sons, Ltd   ISBNs: 0-470-85427-8 (HB); 0-470-85428-6 (PB)

or complex. In this present chapter, the infrared bands of a number of common classes of simple inorganic and coordination compounds are defined. The infrared spectra of metal complexes and minerals are also introduced. The subject of the infrared spectra of inorganic molecules is broad as there is an enormous number of such molecules. While this chapter aims to provide an introduction, there is a number of specialist books on the subject [1–7].

5.2 General Considerations

Simple inorganic compounds, such as NaCl, do not produce any vibrations in the mid-infrared region, although the lattice vibrations of such molecules occur in the far-infrared region. This is why certain simple inorganic compounds, such as NaCl, KBr and ZnSe, are used for infrared windows.

A slightly more complex inorganic, such as $CaCO_3$, contains a complex anion. These anions produce characteristic infrared bands and Table 5.1 summarizes the main bands of some common inorganic ions [8]. The attached cation generally has only a small effect on the wavenumber of the complex anion. The reason for this is the crystal structure formed by such molecules. For example, KNO_2 consists of an ionic lattice with K^+ ions and NO_2^- ions arranged in a regular array. The crystal structure consists of essentially isolated K^+ and NO_2^- ions, and so the infrared bands of the cation and anion are independent. In this example, K^+ is monatomic and produces no vibrations and, hence, no infrared bands. However, heavier cations do cause a band to shift to a lower wavenumber and this effect is more obvious for the bending vibrations observed at lower wavenumbers.

SAQ 5.1

Figure 5.1 below shows the infrared spectrum obtained for a common inorganic compound. Consult Table 5.1 and identify the anion present in this compound.

Table 5.1 Main infrared bands of some common inorganic ions

Ion	Wavenumber (cm^{-1})
CO_3^{2-}	1450–1410, 880–800
SO_4^{2-}	1130–1080, 680–610
NO_3^-	1410–1340, 860–800
PO_4^{3-}	1100–950
SiO_4^{2-}	1100–900
NH_4^+	3335–3030, 1485–1390
MnO_4^-	920–890, 850–840

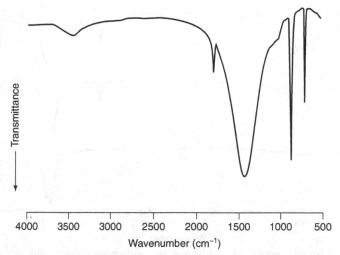

Figure 5.1 Infrared spectrum of an unknown inorganic compound (cf. SAQ 5.1). Used with permission from the *Journal of Chemical Education*, **78**, No. 8, 2001, pp. 1090–1092; copyright ©2001, Division of Chemical Education, Inc.

Several factors have an impact on the appearance of the infrared spectra of inorganic compounds. The crystal form of the compound needs to be considered. Crystalline lattice bands manifest themselves in the far-infrared region and changes to the crystal structure will be observed in the spectra. The consequence is that non-destructive sampling techniques are preferred for such samples. Techniques such as alkali halide discs or mulls can produce pressure-induced shifts in the infrared bands of such materials.

The degree of hydration of an inorganic compound is also a factor when interpreting spectra. The water molecules that are incorporated into the lattice structure of a crystalline compound produce characteristic sharp bands in the 3800–3200 and 1700–1600 cm^{-1} regions, due to O–H stretching and bending, respectively. The lattice environment of the water molecules determines the position of the infrared bands of water and whether they are single or split. The hydroxy stretching bands in the 3800–3200 cm^{-1} range show unique patterns that may be used to characterize the compositions of hydrated inorganic compounds.

SAQ 5.2

Calcium sulfate is commonly found in three forms, i.e. anhydrous (called anhydrite), hemihydrate (plaster of Paris) and dihydrite (gypsum). The mid-infrared spectra of each of these forms are shown below in Figure 5.2. Comment on the differences between these spectra.

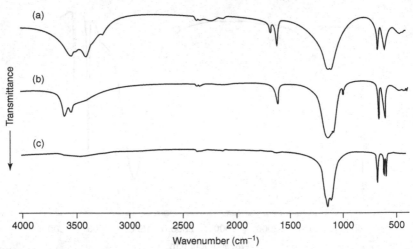

Figure 5.2 Infrared spectra of (a) dihydrate, (b) hemihydrate and (c) anhydrous $CaSO_4$. From Christensen, M. C., *Infrared and Raman Users Group (IRUG) Postprints*, pp. 93–100 (1998), and reproduced by permission of M. C. Christensen.

5.3 Normal Modes of Vibration

In this section, inorganic molecules and ions are considered by their structural types and the vibrational wavenumbers of some common molecules are presented.

Diatomic molecules produce one vibration along the chemical bond. Monatomic ligands, where metals coordinate with atoms such as halogens, H, N or O, produce characteristic infrared bands. These bands are summarized in Table 5.2. Coordination compounds may also contain diatomic ligands, consisting of a metal atom coordinated to molecules such as CO, NO, O_2, N_2, H_2, OH^- or CN^-. Upon coordination, the wavenumber values of such ligands are shifted to lower values.

The normal modes of vibration of linear and bent XY_2 molecules are illustrated in Figure 5.3, with the infrared bands of some common linear and bent triatomic

Table 5.2 Characteristic infrared bands of diatomic inorganic molecules

Wavenumber (cm^{-1})	Assignment[a]
2250–1700	M–H stretching
800–600	M–H bending
750–100	M–X stretching
1010–850	M=O stretching
1020–875	M≡N stretching

[a]M, metal; X, halogen.

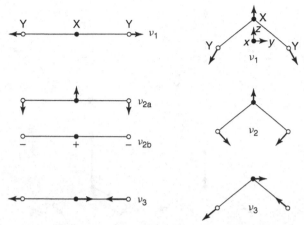

Figure 5.3 Normal modes of vibration of linear and bent XY_2 molecules (+ and − denote vibrations moving upwards and downwards, respectively, in the direction perpendicular to the plane of the paper). From Nakamoto, K., 'Infrared and Raman Spectra of Inorganic and Coordination Compounds', in *Handbook of Vibrational Spectroscopy*, Vol. 3, Chalmers, J. M. and Griffiths, P. R. (Eds), pp. 1872–1892. Copyright 2002. © John Wiley & Sons Limited. Reproduced with permission.

Table 5.3 Characteristic infrared bands (cm^{-1}) of triatomic inorganic molecules

Molecule	ν_1	ν_2	ν_3
Linear			
OCO	1388, 1286	667	2349
HCN	3311	712	2097
NCS^-	2053	486, 471	748
ClCN	714, 784	380	2219
$MgCl_2$	327	249	842
Bent			
H_2O	3657	1595	3756
O_3	1135	716	1089
$SnCl_2$	354	120	334

molecules being given in Table 5.3. Note that some molecules show two bands for ν_1 because of Fermi resonance. Table 5.3 also provides some examples of linear and bent triatomic ligands.

The more common shapes of XY_3 molecules are planar and pyramidal. The normal modes of vibration of planar and pyramidal XY_3 molecules which are

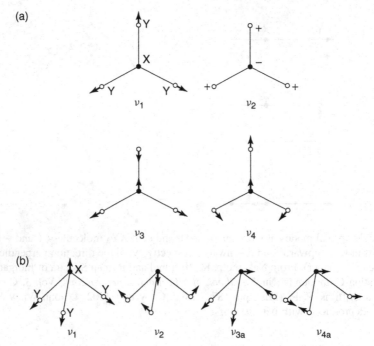

Figure 5.4 Normal modes of vibration of (a) planar and (b) pyramidal XY_3 molecules. From Nakamoto, K., 'Infrared and Raman Spectra of Inorganic and Coordination Compounds', in *Handbook of Vibrational Spectroscopy*, Vol. 3, Chalmers, J. M. and Griffiths, P. R. (Eds), pp. 1872–1892. Copyright 2002. © John Wiley & Sons Limited. Reproduced with permission.

Table 5.4 Characteristic infrared bands (cm^{-1}) of four-atom inorganic molecules

Planar	ν_2	ν_3	ν_4	
BF_3	719	1506	481	
$CaCO_3$	879	1492–1429	706	
KNO_3	828	1370	695	
SO_3	498	1390	530	
Pyramidal	ν_1	ν_2	ν_3	ν_4
PF_3	893	487	858	346
$SO_3{}^{2-}$	967	620	933	469
$ClO_3{}^{-}$	933	608	977	477
$IO_3{}^{-}$	796	348	745	306

'infrared-active' are illustrated in Figure 5.4. The infrared bands of some examples of planar and pyramidal four-atom molecules are shown in Table 5.4. Ligands may also adopt pyramidal and planar structures and the nature of the coordination also affects the resulting infrared bands of these ligands.

The five-atom XY_4 molecules and ligands commonly adopt tetrahedral and square-planar shapes. The normal modes of tetrahedral and square-planar XY_4 are shown in Figure 5.5. Tetrahedral XY_4 molecules show two normal modes that are 'infrared-active', while the square-planar XY_4 molecules show three

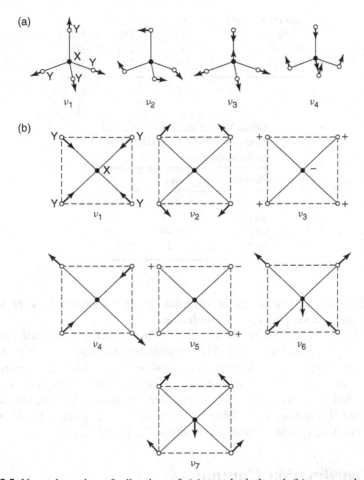

Figure 5.5 Normal modes of vibration of (a) tetrahedral and (b) square-planar XY_4 molecules. From Nakamoto, K., 'Infrared and Raman Spectra of Inorganic and Coordination Compounds', in *Handbook of Vibrational Spectroscopy*, Vol. 3, Chalmers, J. M. and Griffiths, P. R. (Eds), pp. 1872–1892. Copyright 2002. © John Wiley & Sons Limited. Reproduced with permission.

Table 5.5 Characteristic infrared bands (cm^{-1}) of five-atom inorganic molecules

Tetrahedral	ν_3	ν_4	
CH_4	3019	1306	
NH_4^+	3145	1400	
$SnCl_4$	408	126	
PO_4^{3-}	1017	567	
CrO_4^{2-}	890	378	
MnO_4^-	902	386	
Square-planar	ν_3	ν_6	ν_7
XeF_4	291	586	161
$PtCl_4^{2-}$	147	313	165
$PdCl_4^{2-}$	150	321	161

Table 5.6 Characteristic infrared bands (cm^{-1}) of octahedral inorganic molecules

Molecule	ν_3	ν_4
AlF_6^{3-}	568	387
SiF_6^{2-}	741	483
VF_6^{2-}	646	300
$PtCl_6^{2-}$	343	183

'infrared-active' normal modes of vibration. Table 5.5 summarizes the infrared bands of some common five-atom molecules.

An XY_5 molecule may adopt a trigonal bipyramidal, a tetragonal pyramidal or a planar-pentagonal structure. The trigonal bipyramidal, XY_5 (e.g. SF_5^- or BrF_5), shows six normal 'infrared-active' vibrations, while planar-pentagonal XY_5 molecules (e.g. XeF_5^-) show three 'infrared-active' normal vibrations.

Octahedral XY_6 molecules show six normal modes of vibration, but only two of these are 'infrared-active'. Table 5.6 summarizes the infrared bands of some hexahalo molecules which show an octahedral structure.

5.4 Coordination Compounds

Metal complexes or chelates are largely covalent in nature and the spectra of such compounds are dominated by the contribution of the ligand and its coordination chemistry [2]. The ligands may be small species, such as water or ammonium

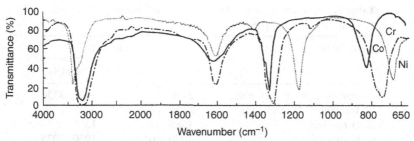

Figure 5.6 Infrared spectra of metal hexamine complexes. From Nakamoto, K., *Infrared and Raman Spectra of Inorganic and Coordination Compounds, Part B, Applications in Coordination, Organometallic and Bioinorganic Chemistry*, 5th Edn. Copyright © (1997 John Wiley & Sons, Inc.). This material is used by permission of John Wiley & Sons, Inc.

molecules, or large complex species, such as porphyrins. Amine complexes have been widely studied and Figure 5.6 illustrates the infrared spectra of the Co(III), Cr(III) and Ni(II) hexamine complexes [9]. The spectra indicate that the type of metal in the complex produces important differences in the infrared bands of each complex. Significant differences in the wavenumber values of each complex are noted for the metal–NH_3 rocking (900–600 cm^{-1}) and NH_3 bending (1400–1100 cm^{-1}) bands. Bands due to N–H stretching and N–H bending are also observed in these spectra in the 3700–2500 and 1750–1500 cm^{-1} regions, respectively.

DQ 5.1

Given that the stability order of the hexamine compounds, the spectra of which are shown in Figure 5.6, is known to be Co(III) > Cr(III) > Ni(II), what do the infrared spectra indicate about the bonding in these compounds?

Answer

The spectra presented in Figure 5.6 show a trend in the wavenumber shifts for the three hexamine complexes; the N–H bands shift to lower wavenumbers from Co to Cr to Ni. This indicates that the N–H bond order (bond strength) decreases as the metal–N bond order increases in the stability order mentioned.

SAQ 5.3

Zinc forms a coordination compound with ethylenediamine (en) via bidentate bridging. The main infrared bands of both en and ZnenCl$_2$ are listed below in Table 5.7 [10]. Compare the spectral bands and comment on the nature of the bonding in the compound.

Table 5.7 Infrared spectral data of ethylenediamine and a zinc(II) ethylenediamine complex (cf. SAQ 5.3)

Assignment	Wavenumber (cm^{-1})	
	en	ZnenCl$_2$
Asymmetric N–H stretching	3440, 3350	3290, 3230
Symmetric N–H stretching	3260	3140
N–H bending	1585	1570
C–N stretching	1090, 1040	1050, 1025

5.5 Isomerism

Linkage and geometrical isomerisms are important issues in coordination chemistry, producing structures with differing properties. Linkage isomerism occurs when a ligand can coordinate to a central metal using either of two atoms within the ligand. The nitrite ion exhibits this type of isomerism. When the nitrite ion is attached to a central metal ion via the nitrogen atom, it is known as a *nitro* ligand. When one of the oxygen atoms is the donor, it is known as a *nitrito* ligand. When NO_2^- bonds undergo bonding through an oxygen bond, one of the NO bonds is 'nearly' a double bond, while the other is a single bond. In the other coordination, both NO bonds are intermediate between single and double bonds. The infrared band of a bond increases as its strength increases and so it would be expected that the wavenumbers of the NO bonds in NO_2^- increase in the order: single-bond NO (in O-bonded) < NO (in N-bonded) < double-bond NO (in O bonded). It is observed that in complexes when NO_2^- is bonded through oxygen, N=O stretching appears in the $1500-1400$ cm^{-1} range while N–O stretching appears at $1100-1000$ cm^{-1}. In complexes in which NO_2^- is bonded through nitrogen, the infrared bands appear at $1340-1300$ cm^{-1} and $1430-1360$ cm^{-1}, which are intermediate values when compared to the oxygen-bonded complex. This demonstrates that it is possible to use infrared spectroscopy to determine whether a nitrite is coordinated and whether it is coordinated through a nitrogen or oxygen atom. Table 5.8 lists the main infrared bands observed for some common ligands capable of forming linkage isomers.

An example of how geometrical isomers may be differentiated by using infrared spectroscopy is the *cis–trans* isomerism exhibited by tetrachlorobis(N,N-dimethylformamide)tin(IV) (abbreviated as SnCl$_4$(DMF)$_2$) [11]. Infrared spectroscopy can be used to determine both the coordination mode of a ligand and the geometrical arrangement of ligands around the metal atom. The mid-infrared spectra of the *cis*- and *trans*-isomers of SnCl$_4$(DMF)$_2$ show carbonyl stretching bands at 1651 and 1655 cm^{-1} for the *cis*- and *trans*-isomers, respectively. The nature of the ligand coordination can be determined by comparing these carbonyl bands with that of the isolated DMF molecules. The infrared spectrum of DMF in

Table 5.8 Infrared bands of some common ligands capable of forming linkage isomers

Ligand	Wavenumber (cm^{-1})
Nitro	1430–1360, 1340–1300
Nitrito	1500–1400, 1100–1000
Thiocyanato (–S–C≡N)	2140–2100, 720–680
Cyanato (–O–C≡N)	2210–2000, 1300–1150
Isothiocyanato (–N=C=S)	2100–2040, 850–800
Isocyanato (–N=C=O)	2250–2150, 1450–1300

CCl_4 shows a C=O stretching band at 1687 cm^{-1}, a notably higher wavenumber. If coordination in the $SnCl_4(DMF)_2$ compound is through the nitrogen atom, the C–O bond order would be increased and the C=O stretching band will shift to a higher wavenumber. In comparison, coordination through the oxygen atom will decrease the C=O stretching wavenumber value. Thus, the positions of the carbonyl bands of *cis*- and *trans*-$SnCl_4(DMF)_2$ indicate that the ligand coordinates via the oxygen atoms. The far-infrared spectra of *trans*- and *cis*-$SnCl_4(DMF)_2$, produced using Nujol mulls between CsI windows, are illustrated in Figure 5.7, and there are important differences in the far-infrared spectra of the isomers of this compound. Both isomers show asymmetric Sn–O stretching at 425 cm^{-1} and a ligand band near 400 cm^{-1}. However, while the *trans*-isomer shows a single band due to Sn–Cl stretching near 340 cm^{-1}, the *cis*-isomer shows four Sn–Cl stretching bands in the 350–290 cm^{-1} range. The *cis*-isomer also shows a symmetric Sn–O stretching band, which is not present in the spectrum of the *trans*-isomer. These notable differences are due to the different symmetries of the molecules [12].

5.6 Metal Carbonyls

A useful band in the infrared spectra of carbonyl ligands in metal complexes is that due to C–O stretching. The latter gives very strong sharp bands which are separated from the bands of other ligands that may be present. The stretching wavenumber for a terminal carbonyl ligand in a complex correlates with the 'electron-richness' of the metal. The band position is determined by the bonding from the d orbitals of the metal into the π^* anti-bonding orbitals of the ligand (known as *backbonding*). The bonding weakens the C–O bond and lowers the wavenumber value from its value in free CO.

DQ 5.2

Account for the variation in the carbonyl stretching bands of the following compounds: CO (2143 cm^{-1}); $[Mn(CO)_6]^+$ (2090 cm^{-1}); $Cr(CO)_6$ (2000 cm^{-1}); $[V(CO)_6]^-$ (1860 cm^{-1}).

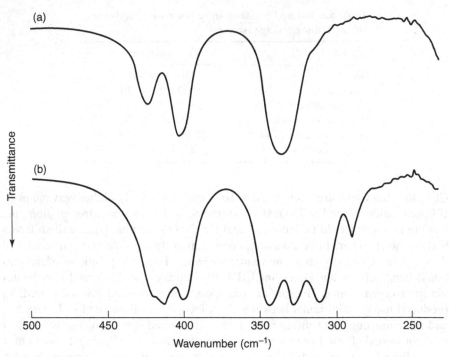

Figure 5.7 Far-infrared spectra of (a) *trans-* and (b) *cis-*SnCl$_4$(DMF)$_2$. Used with permission from the *Journal of Chemical Education*, **71**, No. 8, 1994, pp. 1083–1084; Copyright ©1994, Division of Chemical Education, Inc.

Answer

The C–O stretching wavenumbers are shifted to lower values when there are changes in the extent of backbonding in the compound. Removing positive charge from the metal causes the shift of electrons from the metal to the CO π* orbitals causes the CO wavenumber values to decrease. The highest excess of negative charge on the metal occurs in the [V(CO)$_6$]$^-$ complex and so more backbonding occurs than in the other complexes. The next highest excess of electron density is in Cr(CO)$_6$, and then [Mn(CO)$_6$]$^+$.

Infrared spectroscopy can be readily used to distinguish types of bonding in metal carbonyls [13, 14]. A widely studied case is that of the carbonyl group which occurs in terminal (M–CO) and bridging (e.g. M–CO–M) environments. The bridging CO ligands appear at lower wavenumber values than those of the terminal ligands in complexes with the same metal and similar electron density. Terminal CO bands appear in the broad range from 2130 to 1700 cm^{-1}, while bridging

CO ligand bands appear in the 1900–1780 cm^{-1} range. However, care should be exercised in assigning such carbonyl bands. A band appearing below 1900 cm^{-1} may well be due to a terminal ligand, with a severe reduction of the carbonyl bond strength through d→π* bonding. However, if the complex also shows carbonyl bands well above 1900 cm^{-1}, it may be assumed that such electronic effects are absent and that the lower-wavenumber CO stretching bands can be attributed to bridging ligands.

SAQ 5.4

The infrared spectrum of a $Ru_3(CO)_{12}$ sample shows CO stretching bands at 2060, 2030 and 2010 cm^{-1}, while the spectrum of $Fe_3(CO)_{12}$ shows bands at 2043, 2020, 1997 and 1840 cm^{-1}. What do the spectra illustrate about the bonding in these compounds?

5.7 Organometallic Compounds

Organometallic compounds contain ligands that bond to metal atoms or ions through carbon bonds. The metal–carbon stretching wavenumbers of organometallic compounds are observed in the 600–400 cm^{-1} range, with lighter metals showing bands at higher values. The CH_3 bending bands, arising from metal–CH_3 groups, appear in the 1210–1180 cm^{-1} region in mercury and tin compounds, and at 1170–1150 cm^{-1} in lead compounds. Aromatic organometallic molecules show a strong band near 1430 cm^{-1}, due to benzene ring stretching for metals directly attached to the benzene ring.

5.8 Minerals

Multi-component analysis can be readily applied to the infrared spectra of minerals. The latter contain non-interacting components and so the spectrum of a mineral can be analysed in terms of a linear combination of the spectra of the individual components. However, the spectra of such solids exhibit a marked particle-size dependency. The particle size should be reduced (to 325 mesh) prior to preparation of an alkali halide disc. The pellet preparation involves separate grinding and dispersion steps because minerals tend not to be effectively ground in the presence of an excess of KBr. Figure 5.8 illustrates the analysis of a mineral containing several components. The sample spectrum (a) is shown, as well as the calculated spectrum (b) based on the reference spectra of a variety of standard mineral components. The residual difference spectrum (c) shows that the error between the two spectra is small.

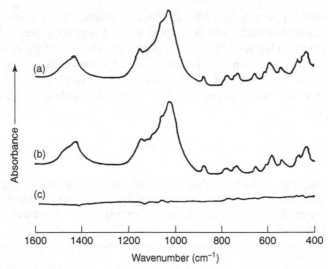

Figure 5.8 Analysis of the infrared spectrum of a mineral. (a) Sample spectrum: 50% albite (k-feldspar); 23% siderite; 17% illite; 10% chlorite. (b) Calculated spectrum: 49.7% feldspar (8.0% albite, 13.2% orthoclase and 28.5% microcline); 25.2% siderite; 19.0% illite; 6.8% chlorite. (c) Residual difference spectrum. From Brown, J. M. and Elliot, J. J., 'The Quantitative Analysis of Complex, Multicomponent Mixtures by FTIR; the Analysis of Minerals and of Interacting Organic Blends', in *Chemical, Biological and Industrial Applications of Infrared Spectroscopy*, Durig, J. R. (Ed.), pp. 111–125. Copyright 1985. © John Wiley & Sons Limited. Reproduced with permission.

Infrared spectroscopy has been frequently used to investigate the structural properties of clay minerals [14–18]. These materials are hydrated aluminium silicates with a layered structure formed by tetrahedral sheets (containing Si(IV)) via shared oxygen atoms. Clay minerals are usually examined by using transmission infrared methods with a KBr disc. Clay minerals may be differentiated by their infrared spectra through a study of the bands due to the O–H and Si–O groups. In the O–H stretching region, $3800–3400$ cm^{-1} for clay minerals, there are a number of bands observed. The inner hydroxyl groups between the tetrahedral and octahedral sheets result in a band near 3620 cm^{-1}. The other three O–H groups at the octahedral surface form weak hydrogen bonds with the oxygens of the Si–O–Si bonds in the next layer and this results in stretching bands at 3669 and 3653 cm^{-1}. Where clays have most of their octahedral sites occupied by divalent central atoms such as Mg(II) or Fe(II), a single band in the O–H stretching region is often observed.

In the $1300–400$ cm^{-1} region, clay minerals show Si–O stretching and bending and O–H bending bands. The shape and position of the bands depend very much on the arrangement within the layers. For example, for kaolinite or dickite, which mainly have Al(III) in the octahedral position, several well-resolved strong bands in the 1120–1000 cm^{-1} region are observed. In comparison, for crysotile, which mainly contains Mg(II) in the octahedral sites, the main Si–O band is observed at 960 cm^{-1}. The O–H bending bands are strongly influenced by the layering in clays. Where the octahedral sheets are mainly occupied by trivalent central atoms, the O–H bending bands occur in the 950–800 cm^{-1} region. Where most of the octahedral sites are occupied by divalent central atoms, the O–H bending bands are shifted to lower wavenumbers in the 700–600 cm^{-1} range.

SAQ 5.5

The mid-infrared spectrum of the clay kaolinite in KBr is illustrated below in Figure 5.9. Assign the infrared bands for kaolinite. What information does this spectrum provide about the structure of this clay?

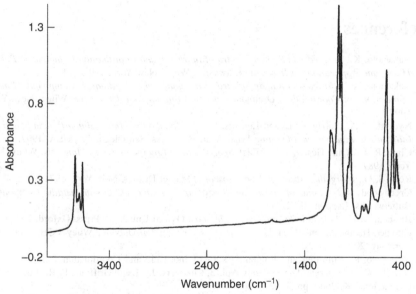

Figure 5.9 Infrared spectrum of kaolinite (cf. SAQ 5.5).

Zeolites are a class of aluminosilicate minerals which are widely used as catalysts and have been well characterized by using infrared spectroscopy [19]. Zeolites possess porous and crystalline structures that are sensitive to excessive pressure, and so a non-destructive sampling technique, such as the use of a diamond ATR cell, is suitable for studying such structures.

Summary

In this chapter, the fundamental features of the infrared spectra of inorganic compounds were introduced. Effects, such as the degree of hydration, on the appearance of infrared spectra were described. The infrared spectra of inorganic molecules are determined by the normal modes of vibration exhibited by such molecules and these were summarized and common examples provided. Infrared spectroscopy is widely used to characterize coordination compounds and the important effects of coordination were introduced. Isomerism in coordination compounds may also be characterized by using infrared techniques and examples were provided. Infrared spectroscopy is extensively used to characterize metal carbonyl compounds and examples of how the bonding in such compounds may be understood were provided. The main infrared bands associated with organometallic compounds were also introduced. Finally, examples of how infrared spectroscopy may be employed to understand the structures of mineral compounds were included.

References

1. Nakamoto, K., *Infrared and Raman Spectra of Inorganic and Coordination Compounds, Part A, Theory and Applications in Inorganic Chemistry*, Wiley, New York, 1997.
2. Nakamoto, K., *Infrared and Raman Spectra of Inorganic and Coordination Compounds, Part B, Applications in Coordination, Organometallic and Bioinorganic Chemistry*, Wiley, New York, 1997.
3. Nyquist, R. A., Putzig, C. L. and Leugers, M. A., *Handbook of Infrared and Raman Spectra of Inorganic Compounds and Organic Salts*, Academic Press, San Diego, CA, USA, 1997.
4. Clark, R. J. H. and Hester, R. E. (Eds), *Spectroscopy of Inorganic Based Materials*, Wiley, New York, 1987.
5. Ross, S. D., *Inorganic Vibrational Spectroscopy*, Marcel Dekker, New York, 1971.
6. Greenwood, N. N., *Index of Vibrational Spectra of Inorganic and Organometallic Compounds*, Butterworths, London, 1972.
7. Brisdon, A. K., *Inorganic Spectroscopic Methods*, Oxford University Press, Oxford, UK, 1998.
8. Gunzler, H. and Gremlich, H.-U., *IR Spectroscopy: An Introduction*, Wiley-VCH, Weinheim, Germany, 2002.
9. Nakamoto, K., 'Infrared and Raman Spectra of Inorganic and Coordination Compounds', in *Handbook of Vibrational Spectroscopy*, Vol. 3, Chalmers, J. M. and Griffiths, P. R. (Eds), Wiley, Chichester, UK, 2002, pp. 1872–1892.
10. Ahuja, I. S. and Tripathi, S., *J. Chem. Edu.*, **68**, 681–682 (1992).
11. Tudela, D., *J. Chem. Edu.*, **71**, 1083–1084 (1994).
12. Ebsworth, E. A. V., Rankin, D. W. H. and Cradock, S., *Structural Methods in Inorganic Chemistry*, Blackwell, Oxford, 1987.
13. Cotton, F. A., Wilkinson, G., Murillo, C. A. and Bochmann, M., *Advanced Inorganic Chemistry*, 6th Edn, Wiley, New York, 1999.
14. Madejova, J., *Vibr. Spectrosc.*, **31**, 1–10 (2003).
15. Farmer, V. C. (Ed.), *Infrared Spectra of Minerals*, Mineralogical Society, London, 1974.
16. Wilson, M. J. (Ed.), *Clay Mineralogy: Spectroscopic and Chemical Determinative Methods*, Chapman and Hall, London, 1994.

17. Gadsden, J. A., *Infrared Spectra of Minerals and Related Inorganic Compounds*, Butterworths, London, 1975.

18. Busca, G. and Resini, C., 'Vibrational Spectroscopy for the Analysis of Geological and Inorganic Materials', in *Encyclopedia of Analytical Chemistry*, Vol. 12, Meyers, R. A. (Ed.), Wiley, Chichester, UK, 2000, pp. 10984–11020.

19. Zecchina, A., Spoto, G. and Bordiga, S., 'Vibrational Spectroscopy of Zeolites', in *Handbook of Vibrational Spectroscopy*, Vol. 4, Chalmers, J. M. and Griffiths, P. R. (Eds), Wiley, Chichester, UK, 2002, pp. 3042–3071.

Chapter 6
Polymers

Learning Objectives

- To identify and characterize homopolymers, copolymers and polymer blends by using infrared spectroscopy.
- To use infrared spectroscopy to carry out quantitative analysis of additives or contaminants in polymers.
- To understand how infrared spectroscopy may be used to monitor polymerization reactions.
- To use infrared spectroscopy to characterize the structural properties of polymers, including tacticity, branching, crystallinity, hydrogen bonding and orientation.
- To use infrared spectroscopy to investigate the surface properties of polymers.
- To understand how polymer degradation may be monitored by using infrared spectroscopic techniques.

6.1 Introduction

Polymers play an enormously important role in modern society; these materials are fundamental to most aspects of modern life, such as building, communication, transportation, clothing and packaging. Thus, an understanding of the structures and properties of polymers is vital. Polymers are large molecules, generally consisting of a large number of small-component molecules known as monomers. Many polymers are synthesized from their constituent monomers via a polymerization

---

Infrared Spectroscopy: Fundamentals and Applications   B. Stuart
© 2004 John Wiley & Sons, Ltd   ISBNs: 0-470-85427-8 (HB); 0-470-85428-6 (PB)

process. Most commercial polymers are based on covalent compounds of carbon, but synthetic polymers may also be based on inorganic atoms such as silicon.

Infrared spectroscopy is a popular method for identifying polymers [1–7]. In this present chapter, the application of infrared spectroscopy to the study of polymeric systems will be introduced. Infrared spectroscopy may be used to identify the composition of polymers, to monitor polymerization processes, to characterize polymer structure, to examine polymer surfaces and to investigate polymer degradation processes.

6.2 Identification

There are a number of methods available for examining polymer samples [7, 8]. If the polymer is a *thermoplastic*, it can be softened by warming and formed in a hydraulic press into a thin film. Alternatively, the polymer may be dissolved in a volatile solvent and the solution allowed to evaporate to give a thin film on an alkali halide plate. Some polymers, such as cross-linked synthetic rubbers, can be *microtomed* (cut into thin slices with a blade). A solution in a suitable solvent is also a possibility. If the polymer is a surface coating, reflectance techniques may be used.

As most polymers are organically based, the spectral assignments made for organic molecules in Chapter 4 are helpful when interpreting the infrared spectra of polymers. A useful correlation table for polymers is shown in Figure 6.1.

Figure 6.2 illustrates the infrared spectrum of poly(methyl methacrylate) (PMMA). The structure of PMMA is $-[-CH_2-C(CH_3)(COOCH_3)-]_n-$. The infrared spectrum of PMMA results from the group frequencies of the C–C and C–H groups of the backbone chain, the C–C, C=O and C–O units of the ester group and the C–H units of the methyl substituent. The infrared assignments for this spectrum are listed in Table 6.1.

Table 6.1 Major infrared bands of poly(methyl methacrylate)

Wavenumber (cm^{-1})	Assignment
2992	O–CH$_3$, C–H stretching
2948	C–CH$_3$, C–H stretching
1729	C=O stretching
1485	CH$_2$ bending
1450,1434	O–CH$_3$ bending
1382,1337	C–CH$_3$ bending
1265,1238	C–C–O stretching
1189,1170	C–O–C bending
1145	CH$_2$ bending
962	C–CH$_3$ bending

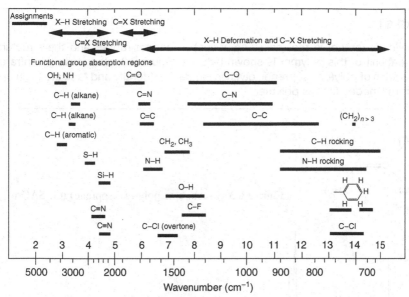

Figure 6.1 Correlation table for the infrared bands of polymers. Reprinted from *Polymer Synthesis and Characterization: A Laboratory Manual*, Sandler, S. R., Karo, W., Bonesteel, J. and Pearce, E. M., Figure 14.1, p. 99, Copyright (1998), with permission from Elsevier.

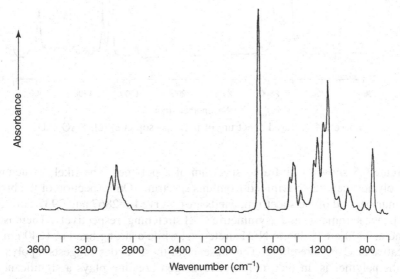

Figure 6.2 Infrared spectrum of poly(methyl methacrylate). From Stuart, B., *Modern Infrared Spectroscopy*, ACOL Series, Wiley, Chichester, UK, 1996. © University of Greenwich, and reproduced by permission of the University of Greenwich.

SAQ 6.1

Poly(*cis*-isoprene) is the main component of natural rubber and the structural repeat unit of this polymer is shown below in Figure 6.3, along with the infrared spectrum of poly(*cis*-isoprene), given in Figure 6.4. Identify and tabulate the major infrared modes for this polymer.

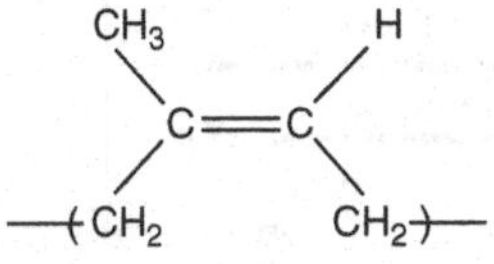

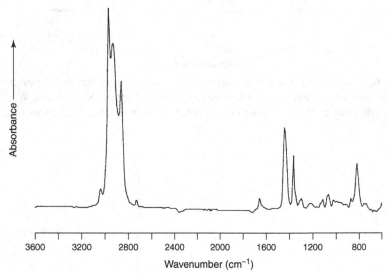

Figure 6.3 Structure of poly(*cis*-isoprene) (cf. SAQ 6.1).

Figure 6.4 Infrared spectrum of poly(*cis*-isoprene) (cf. SAQ 6.1).

Figure 6.5 shows an infrared spectrum of a polymer. The likely structure of this polymer may be determined from this spectrum. On inspection of the higher-wavenumber end of the spectrum, bands are observed at 2867 and 2937 cm^{-1}, due to aliphatic symmetric and asymmetric C–H stretching, respectively. There is also a band at 3300 cm^{-1} due to N–H stretching, while a strong band at 1640 cm^{-1} is indicative of C=O stretching. The presence of these modes suggests a polyamide and the polymer is, in fact, nylon 6. Hydrogen bonding plays a significant role in the spectra of nylons. The N–H stretching mode is due to hydrogen-bonded groups, while the broad shoulder which appears at around 3450 cm^{-1} is due to non-hydrogen-bonded N–H bonds.

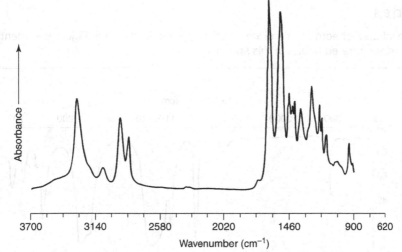

Figure 6.5 Infrared spectrum of an unknown polymer. From *Polymer Analysis*, AnTS Series, Stuart, B. H. Copyright 2002. © John Wiley & Sons Limited. Reproduced with permission.

SAQ 6.2

Figure 6.6 shows the infrared spectrum of an unknown polymer film. What type of polymer was used to produced this film?

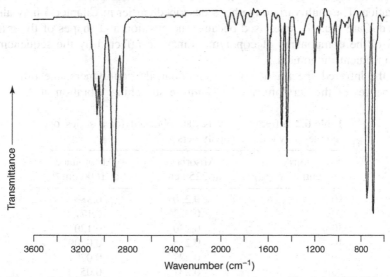

Figure 6.6 Infrared spectrum of an unknown polymer film (cf. SAQ 6.2).

SAQ 6.3

The infrared spectrum of a silicon-based polymer is shown in Figure 6.7. Identify the major infrared bands in this spectrum.

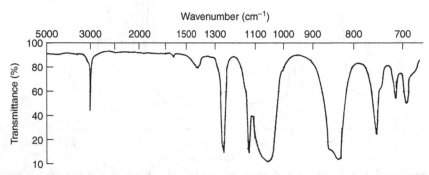

Figure 6.7 Infrared spectrum of an unknown silicon-based polymer (cf. SAQ 6.3).

Copolymers are comprised of chains containing two or more different types of monomers. The composition of copolymers may be quantitatively determined by using infrared spectroscopy [3, 8]. Distinctive representative modes for the polymers may be identified. For example, in the case of vinyl chloride–vinyl acetate copolymers, the ratio of the absorbance of the acetate mode at 1740 cm^{-1} to that of the vinyl chloride methylene bending mode at 1430 cm^{-1} can be used for quantitative analysis. Copolymers of known composition may be used for calibration. The multivariate methods described earlier in Chapter 3 may also be applied. Care must be exercised because the position and shapes of the infrared bands of the components of copolymers may be affected by the sequencing of the constituent monomers.

In the infrared spectra of styrene–acrylonitrile copolymers, the ratio of the absorbances of the acrylonitrile C≡N nitrile stretching vibration at 2250 cm^{-1}

Table 6.2 Infrared analysis data obtained for a series of styrene–acrylonitrile copolymers

Acrylonitrile concentration (%)	Absorbance at 2250 cm^{-1}	Absorbance at 1600 cm^{-1}
10	0.230	0.383
20	0.223	0.177
30	0.230	0.120
40	0.235	0.0909
50	0.227	0.0701
60	0.231	0.0592

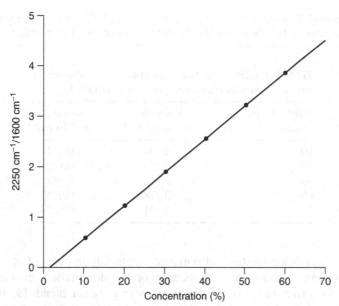

Figure 6.8 Calibration plot for styrene–acrylonitrile copolymers. From *Polymer Analysis*, AnTS Series, Stuart, B. H. Copyright 2002. © John Wiley & Sons Limited. Reproduced with permission.

and the styrene ring stretching vibration at 1600 cm^{-1} may be employed as a measure of the composition. Infrared analysis of a number of styrene–acrylonitrile copolymers of known composition yielded the results listed in Table 6.2. A plot of the absorbance ratio, 2250 cm^{-1}/1600 cm^{-1}, as a function of concentration, is linear and can be used as a calibration plot (Figure 6.8). The infrared spectrum of a sample of poly(styrene-*co*-acrylonitrile) of unknown composition was recorded. The absorbance values at 2250 and 1600 cm^{-1} in this spectrum were 0.205 and 0.121, respectively. The appropriate absorbance ratio for the unknown sample is 0.205/0.121 = 1.70 and consultation with the calibration graph gives a concentration of 26.6% acrylonitrile for the copolymer. Thus, the composition of the copolymer is 73.4% styrene/26.6% acrylonitrile.

SAQ 6.4

In the infrared spectra of ethylene–vinyl acetate copolymer films, the ratio of the absorbances of the vinyl acetate C–O stretching vibration at 1020 cm^{-1} and the ethylene CH$_2$ rocking vibration at 720 cm^{-1} may be employed as a measure of the composition. Infrared analysis of a number of ethylene–vinyl acetate copolymers of known composition produced the results listed in Table 6.3. The infrared spectrum of a sample of poly(ethylene-*co*-vinyl acetate) of unknown composition was recorded. The absorbance values at 1020 and 720 cm^{-1} in this spectrum were 0.301 and 0.197, respectively. Estimate the composition of this sample.

Table 6.3 Infrared analysis data obtained for a number of ethylene–vinyl acetate copolymers (cf. SAQ 6.4)

Vinyl acetate concentration (%)	Absorbance at 1020 cm^{-1}	Absorbance at 720 cm^{-1}
10	0.204	0.234
20	0.253	0.138
30	0.388	0.142
40	0.278	0.0757
50	0.301	0.0654

Polymer blends are mixtures of polymeric materials and consist of at least two polymers or copolymers. Infrared spectroscopy is now quite a commonly used technique for examination of the interactions in polymer blends [9, 10]. If two polymers are immiscible, the infrared spectrum should be the sum of the spectra of the two components. Phase-separation implies that the component polymers in the blend will have an environment similar to the pure polymers. If the polymers are miscible, there is the possibility of chemical interactions between the individual polymer chains. Such interactions may lead to differences between the spectra of the polymers in the blend and the pure components. Generally, wavenumber shifts and band broadening are taken as evidence of chemical interactions between the components in a blend and are indicative of miscibility.

FTIR spectroscopy has been used to investigate the miscibility of certain poly (vinyl phenol) (PVPh) ([–CH$_2$–CH(C$_6$H$_4$)OH–]$_n$–) blends [11]. PVPh has been blended in one case with poly(ethylene oxide) (PEO) (–[–CH$_2$–CH$_2$–O–]$_n$–), and in the other case with poly(vinyl isobutylether) (PVIE) (–[–CH$_2$–CH(OC$_4$H$_9$)–]$_n$–). Figures 6.9 and 6.10 show the O–H stretching regions of the infrared spectra of these blends for a number of compositions. Figure 6.9 shows the O–H stretching region of PVPh of the PVPh/PEO blend. The band at 3360 cm^{-1} shifts to a lower wavenumber near 3200 cm^{-1} with increasing concentration of PEO. This change reflects an increase in hydrogen bonding between the PVPh hydroxyl groups and the PEO ether oxygens. Such an interaction indicates that the PVPh/PEO blend is miscible. Figure 6.10 illustrates the O–H stretching region of the PVPh/PVIE blend. No changes to the hydroxyl group of PVPh are observed at any concentration, hence indicating that there are no interactions between these two polymers, and so this blend in immiscible.

Polymers are commonly used for the manufacture of fibres and acrylic-based polymer fibres are widely used in synthetic textiles. Acrylonitrile copolymerized with monomers such as vinyl acetate, methyl acrylate and methyl methacrylate are

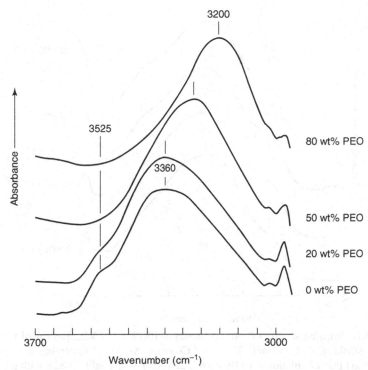

Figure 6.9 Infrared spectra of various blends of PVPh and PEO. Reprinted from *Polymer*, **26**, Moskala, E. J., Varnell, D. F. and Coleman, M. M., 'Concerning the miscibility of poly(vinyl phenol) blends – FTIR study', 228–234, Copyright (1985), with permission from Elsevier.

common and there is a wide variety of combinations used. For forensic purposes, it is important to be able to differentiate synthetic fibres. Infrared microscopy may be used to identify the chemical composition of single fibres and is a valuable forensic tool [12–15].

DQ 6.1

Nylons are also widely used in the manufacture of fibres. However, there is a variety of commercial nylons available, each with slightly different structures, but all containing an amide group and methylene chains. Suppose that two textile fibres are examined by using infrared microscopy, where one is known to be nylon 6 and the other nylon 6,6. Given that nylon 6 has the structure $-[-NH-(CH_2)_5-CO-]_n-$ and that nylon 6,6 has the structure $-[-NH-(CH_2)_6-NH-CO-(CH_2)_4-CO-]_n-$, how could the infrared spectra obtained for each of the fibres be used to identify which fibre is nylon 6 and which is nylon 6,6?

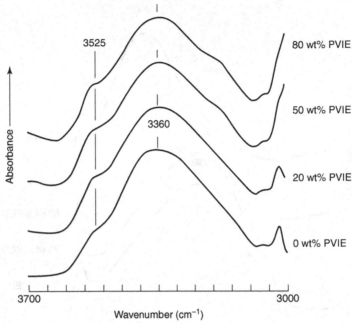

Figure 6.10 Infrared spectra of various blends of PVPh and PVIE. Reprinted from *Polymer*, **26**, Moskala, E. J., Varnell, D. F. and Coleman, M. M., 'Concerning the miscibility of poly(vinyl phenol) blends – FTIR study', 228–234, Copyright (1985), with permission from Elsevier.

Answer

As nylons are polyamides, infrared bands due to the amide group will be significant in their spectra. The relative positions of the N–H and C=O groups in the respective nylon structures will affect the nature of the hydrogen bonding in each of these molecules, and hence will affect the positions of the amide bands in the spectrum. When the spectra of nylon 6 and nylon 6,6 are compared, differences in the 1450–850 cm^{-1} region are detectable. In particular, the amide II bands near 1270 cm^{-1} are shifted by more that 10 cm^{-1} in the spectra.

Infrared spectroscopy is also readily employed for the quantitative analysis of additives or contaminants in polymers. For example, the amount of dioctyl phthalate (DOP), a common plasticizer in poly(vinyl chloride) (PVC), may be determined by using the absorbance ratio of a 1460 cm^{-1} band in the infrared spectrum of DOP and a 1425 cm^{-1} band in the PVC spectrum. A calibration curve can be set up by using the infrared spectra of a series of PVC samples containing known values of DOP content: the absorbance ratio, 1460 cm^{-1} DOP/1425 cm^{-1} PVC, versus concentration is then plotted.

SAQ 6.5

The infrared spectra of polyurethanes may be used to analyse for the presence of residual isocyanate in the polymers. It is important to quantify the amount of isocyanate as the presence of this reagent affects the curing process. Four samples of polyurethane prepolymers, containing known amounts of isocyanate, dissolved in chloroform have been studied by using infrared spectroscopy. The samples were examined by using a CaF_2 transmission liquid cell and the spectral contributions of chloroform were removed by spectral subtraction. The absorbance of the N=C=O stretching band of isocyanate may be used for quantitative analysis and Table 6.4 lists the absorbance value of this band for each sample. Another sample of polyurethane prepolymer, studied under the same conditions, containing an unknown quantity of isocyanate, was also examined by using infrared spectroscopy. Estimate the amount of isocyanate present in the unknown sample.

Table 6.4 Infrared analysis data obtained for isocyanate in a polyurethane prepolymer (cf. SAQ 6.5)

Isocyanate (%)	Absorbance at 2270 cm^{-1}
3	0.094
6	0.194
9	0.304
12	0.410
Unknown	0.254

6.3 Polymerization

As the infrared spectra of monomers are different to those of the corresponding polymers, it is possible to use infrared spectroscopy to monitor polymerization reactions. The technique has also been widely applied to the study of curing [6, 16]. For example, the extent of cross-linking of epoxy resins with amines can be examined by using the C–O stretching and C–H stretching bands because the cross-linking process involves opening of the epoxy ring. For instance, the absorbances of the 912 and 3226 cm^{-1} bands may be measured as a function of time to follow the reaction.

Infrared spectroscopy has been successfully applied to the study of polyure-thane reactions. Attenuated total reflectance (ATR) spectroscopy is particularly useful for monitoring the infrared spectra of a reaction mixture over short time intervals and for characterizing the reaction progress [2]. For instance, the rate of consumption of free isocyanate groups and the competitive formation of urethane, isocyanurate and urea linkages can be deduced from the decrease of the N=C=O stretching mode at 2275 cm^{-1}, and the increase of the C=O stretching modes at 1725, 1710 and 1640 cm^{-1}, respectively.

6.4 Structure

The physical properties of polymers are affected by the structures of the molecular chains. Depending on the nature of the polymerization, certain polymers are able to form different configurational isomers. Infrared spectroscopy has been applied to studies of the tacticity of a number of vinyl polymers [1, 6, 17]. In stereoisomers, the atoms are linked in the same order in a *head-to-tail configuration*, but differ in their spatial arrangement. In particular for asymmetric monomers, the orientation of each monomer adding to the growing chain is known as *tacticity*. There are three stereoisomers observed for vinyl polymers: *isotactic*, where all of the substituent (R) groups are situated on the same side of the polymer chain (Figure 6.11(a)); *syndiotactic*, where the substituent groups are located on alternate sides of the polymer chain (Figure 6.11(b)); *atactic*, where the substituent groups are randomly positioned along the polymer chain (Figure 6.11(c)). There is a relationship between stereoregularity and the presence of regular chains and this leads to the appearance of infrared bands due to regular chains. The infrared spectra of the isotactic, syndiotactic and atactic forms of polypropylene (PP) display characteristic differences, as shown in Figure 6.12. The absorbance values at 970 and 1460 cm^{-1} do not depend upon the tacticity,

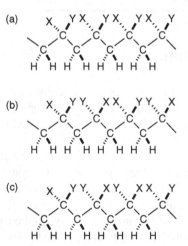

Figure 6.11 Stereoisomers observed for asymmetric vinyl polymers: (a) isotactic; (b) syndiotactic; (c) atactic. From *Polymer Analysis*, AnTS Series, Stuart, B. H. Copyright 2002, © John Wiley & Sons Limited. Reproduced with permission.

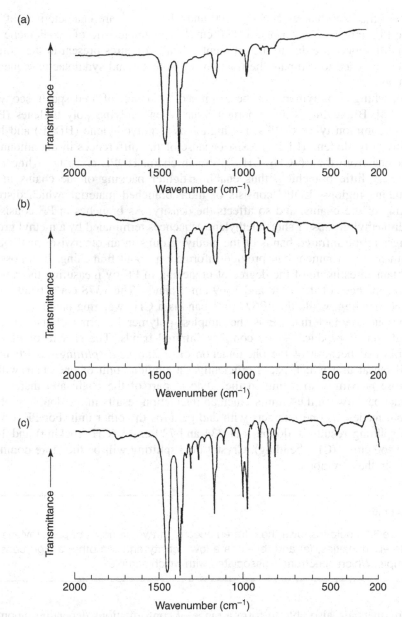

Figure 6.12 Infrared spectra of polypropylene stereoisomers: (a) atactic; (b) syndiotactic; (c) isotactic. From Klopffer, W., *Introduction to Polymer Spectroscopy*, Figure 7.9, p. 91. © Springer-Verlag, 1984. Reproduced by permission of Springer-Verlag GmbH & Co. KG.

whereas the absorbances at 840, 1000 and 1170 cm^{-1} are characteristic of iso-tactic PP, and the absorbance at 870 cm^{-1} is characteristic of syndiotactic PP. Such differences are due to the different helical structures present in the isomers and can be used to estimate the fractions of isotactic and syndiotactic sequences in samples.

Branching in polymers can be examined by using infrared spectroscopy [1, 5, 6, 18]. Branching is an important issue when studying polyethylenes (PEs) as two common types of PEs, i.e. high-density polyethylene (HDPE) and low-density polyethylene (LDPE), exist because of the differences in the amount of branching from the $-(-CH_2-CH_2-)_n-$ main chain. HDPE consists of linear PE with very little branching, thus enabling tighter packing of the chains in the crystalline regions. LDPE consists of more branched material, which disrupts packing of the chains, and so affects the density. As branches in PE consist of either methyl groups or short methylene sequences terminated by a methyl group, changes in the infrared bands of the methyl groups in an otherwise 'methylene-dominant' environment can provide information about branching. It is possible to obtain an estimate of the degree of branching in PE by measuring the ratio of the absorbances of the 1378 and 1369 cm^{-1} bands. The 1378 cm^{-1} band is due to CH_3 bending, while the 1369 cm^{-1} band is a CH_2 wagging band.

Despite the fact that PE is the simplest polymer in terms of its structural repeat unit, it produces some complex infrared bands. The spectra of PEs are complicated because of the phenomenon of *crystal field splitting*. As PE has a small repeat unit which packs efficiently, an ethylene unit of one chain will be in close proximity to a unit in the adjacent part of the chain and there is an interaction between these units. Such an interaction results in the doubling of the normal modes, as one interacts with and perturbs the other unit. For PE, crystal field splitting results in doublets at 734 and 720 cm^{-1} (CH_2 rocking) and 1475 and 1460 cm^{-1} (CH_2 bending). Crystal field splitting will be the more dominant factor in these regions.

SAQ 6.6

Figure 6.13 below shows the infrared spectra of two samples of polyethylene of different densities, (a) and (b) – one a low-density and the other a high-density sample. Which spectrum is associated with which sample?

Polymers are also able to form a range of conformations depending upon the backbone structure. One of the more fundamental aspects of thermoplastic polymers is their ability to form crystalline states and the presence of such regions has a significant effect on the properties of these materials. Infrared spectroscopy provides a suitable method for studying the presence of crystalline regions in polymers, as infrared modes are sensitive to changes in bond angles [1, 5, 6, 9].

(a)

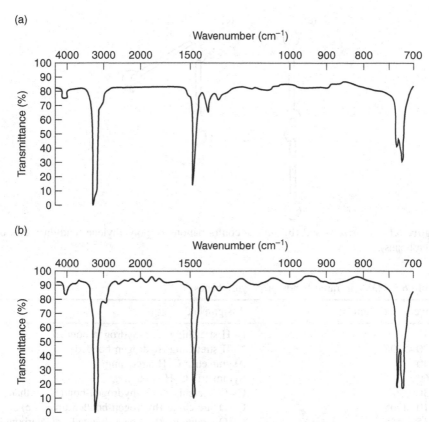

(b)

Figure 6.13 Infrared spectra of samples of LDPE and HDPE (cf. SAQ 6.6).

The most widely studied polymer is polyethylene (PE), which may adopt *trans*- or *gauche*-conformations, as illustrated in Figure 6.14. The crystalline regions of this polymer consist of molecules in a *trans*-conformation and a 622 cm^{-1} band due to CH$_2$ groups is representative of a *trans–trans* structure. The amorphous regions of PE show *trans–gauche* and *gauche–gauche* structures, while methylene wagging bands at 1303, 1353 and 1369 cm^{-1} can be used to determine the concentrations of these structures. The use of infrared bands due to the crystallizable conformers is aided by the observation that these bands become much sharper or narrower when crystallization occurs.

Hydrogen bonding is readily examined by using infrared spectroscopy, as described earlier in Chapter 3, and is of interest in polymer blends where it may be used to understand polymer compatibility [5, 9]. Hydrogen bonding is also an issue for polyurethanes (PUs), which have the general structure, –CO–O–R–O–CO–NH–R′–NH–. The major infrared modes common to PUs are listed in Table 6.5. PUs are extensively hydrogen bonded, with the proton donor

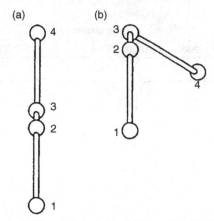

Figure 6.14 (a) *Trans-* and (b) *gauche*-conformations of polyethylene (end-on views of three bonds).

Table 6.5 Major infrared bands of polyurethanes

Wavenumber (cm^{-1})	Assignment
3445	N–H stretching (non-hydrogen bonded)
3320–3305	N–H stretching (hydrogen bonded)
2940	Asymmetric C–H stretching
2860	Symmetric C–H stretching
1730	C=O stretching (non-hydrogen-bonded urethane)
1710–1705	C=O stretching (hydrogen-bonded urethane)
1645–1635	C=O stretching (hydrogen-bonded urea carbonyl)

being the N–H group of the urethane linkage. The hydrogen-bond acceptor may be in either the *hard* segment (the carbonyl of the urethane group) or in the *soft* segment (an ester carbonyl or ester oxygen). The relative amounts of the two types of hydrogen bonds are determined by the degree of microphase separation. An increase in this separation favours the inter-urethane hydrogen bonds. The degree of hydrogen bonding of the N–H groups can be studied by examining the N–H stretching region of the spectrum. The presence of a shoulder in the vicinity of 3450 cm^{-1} is indicative of free N–H groups.

DQ 6.2

Hydrogen bonding plays a fundamental role in the structural and physical properties of nylons and is the most significant type of intermolecular interaction that influences the infrared spectrum of this polymer. The temperature dependence of the infrared spectra of nylons was investigated and the infrared spectra of nylon 6 at 25 and 235°C are shown below in Figure 6.15. What do the differences in these spectra say about the structure of nylon 6 at higher temperatures?

Answer

The main changes observed in the spectra with increasing temperature are associated with the N–H and C=O stretching modes in the 3450–3300 cm^{-1} and 1700–1600 cm^{-1} ranges, respectively. The changes are due to a breakdown of the hydrogen bonds which occur between adjacent chains.

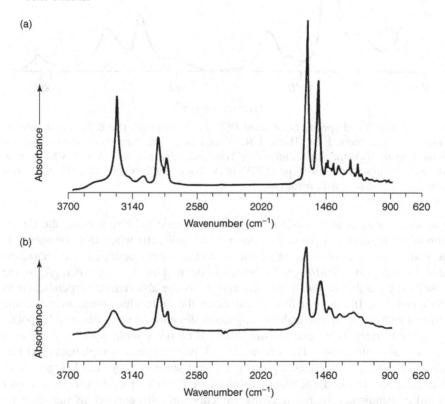

Figure 6.15 Infrared spectra of nylon 6 obtained at temperatures of (a) 25°C and (b) 235°C (cf. DQ 6.2). From *Polymer Analysis*, AnTS Series, Stuart, B. H. Copyright 2002. © John Wiley & Sons Limited. Reproduced with permission.

Infrared spectroscopy is particularly applicable to the study of orientation in polymers [1, 6, 8]. Orientation can be observed in infrared spectroscopy because infrared absorbance is due to the interaction between the electric field vector and the molecular dipole transition moments due to molecular vibrations.

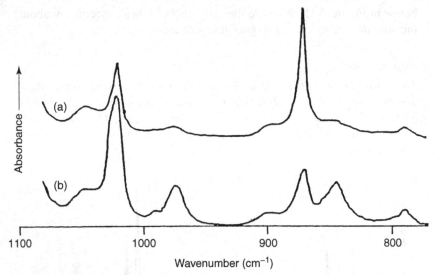

Figure 6.16 Infrared spectra of oriented PET: (a) perpendicular to draw; (b) parallel to draw. From Chalmers, J. M., Hannah, R. W. and Mayo, D., 'Spectra–Structure Correlations: Polymer Spectra', in *Handbook of Vibrational Spectroscopy*, Vol. 3, Chalmers, J. M. and Griffiths, P. R. (Eds), pp. 1893–1918. Copyright 2002. © John Wiley & Sons Limited. Reproduced with permission.

The absorbance is at a maximum when the electric field vector and the dipole transition moment are parallel to each other, and zero when the orientation is perpendicular. The orientation of the molecular components can be characterized by using the *dichroic ratio*, which is defined as $A_{\parallel}/A_{\perp}$, where $A_{\parallel}$ is the absorbance parallel to the chain axis and $A_{\perp}$ is the absorbance perpendicular to the chain axis. If a film or fibre is extended, the molecular groups may become aligned within the sample and this alignment direction may be detected by polarizing the infrared beam and causing it to fall on the sample with the draw axes in particular directions. The orientation of poly(ethylene terephthalate) (PET) has been well characterized by using infrared spectroscopy [1, 8]. Figure 6.16 illustrates the infrared spectra of oriented PET where the electric field vector is (a) perpendicular to the drawing direction and (b) parallel to the drawing direction. These spectra clearly show that the PET spectrum is sensitive to the orientation of the sample, as the differing alignment of the molecules results in changes in the intensities of a number of infrared modes.

6.5 Surfaces

The surface properties of a polymer are of importance in many of the applications in which these materials are employed, including adhesives, lacquers, metallization and composites. Infrared spectroscopic methods, such as attenuated total reflectance, diffuse reflectance, specular reflectance and photoacoustic spectroscopies, allow the surface properties in the order of microns to be characterized.

The ATR spectrum of a nylon 6 film is shown in Figure 6.17. It has been shown that the sampling depth for polymeric materials is about three times d_p [2]. The sampling depths for a polymer which has a refractive index of 1.5, examined by using a 45° ZnSe ATR element, have been calculated at various wavenumbers and are listed in Table 6.6. The sampling depth in this case is typically 2–6 μm.

Infrared spectroscopy is an established procedure for the forensic analysis of pressure-sensitive adhesive tapes [13, 19–21]. ATR spectroscopy is a often

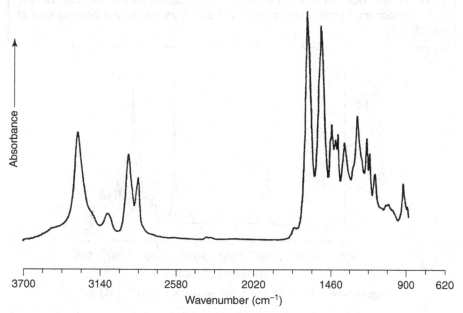

Figure 6.17 ATR spectrum of a nylon 6 film. From *Polymer Analysis*, AnTS Series, Stuart, B. H. Copyright 2002. © John Wiley & Sons Limited. Reproduced with permission.

Table 6.6 The sampling depths for a polymer using a 45° ZnSe ATR element

Wavenumber (cm⁻¹)	Sampling depth (μm)
900	6.3
1600	3.5
3000	1.9

chosen technique as it may be used to obtain the spectra of the surfaces of both the adhesive and backing sides of a tape sample. 'Micro-ATR' techniques are of particular use for such applications, as tape evidence submitted for examination may be soiled and only very small regions may be suitable for analysis. A range of polymers are used for the production of adhesive tapes – acrylate-based adhesives, polyethylene backing (duct tape), poly(vinyl chloride) backing (electrical tape), polypropylene backing (packaging tape) and cellulose acetate backing (office tape).

SAQ 6.7

Figure 6.18 below illustrates the ATR spectrum obtained for a pressure-sensitive adhesive tape. Use this spectrum to identify the material used to manufacture the tape component. Is this a spectrum of the adhesive side or the backing side of the tape?

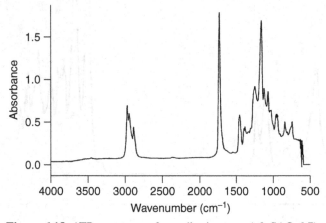

Figure 6.18 ATR spectrum of an adhesive tape (cf. SAQ 6.7).

6.6 Degradation

The degradation and weathering of polymers affect the appearance and the physical properties of these materials, with some common effects being discoloration and embrittlement. Under extreme conditions, the release of volatile products or burning may occur. Infrared spectroscopy may be used to elucidate the degradation mechanisms [5, 22] of polymers by identifying and quantifying the degradation products.

Significant degradation mechanisms for polymers include photo-oxidation and thermo-oxidation. Such mechanisms result in the formation of carbonylated and hydroxylated compounds, which may be identified by examination of the $1900-1500$ cm^{-1} and $3800-3100$ cm^{-1} regions, respectively, in the infrared spectra. Given that there may be a complex mixture of oxidation products, this will result in a complex infrared absorption band. However, the oxidation products may be identified by treating the oxidized polymer samples with a reactive gas, such as SF$_4$ or NH$_3$. Such a derivatization process selectively converts the oxidation products and there will be a subsequent modification of the overlapped infrared bands.

The role of infrared spectroscopy in polymer degradation is illustrated by its application to thermo- and photo-oxidized polyethylenes [1, 8]. During the thermal-oxidation process of PE, a range of carbonyl-containing compounds is formed. These decomposition products give rise to a broad C=O stretching band at about 1725 cm^{-1}, consisting of a number of overlapping component bands. When the oxidized samples are treated with an alkali, a shoulder at 1715 cm^{-1} disappears and is replaced by a distinctive peak near 1610 cm^{-1}. This band is due to C=O stretching of the COO$^-$ ion of a salt, indicating that the shoulder at 1715 cm^{-1} is characteristic of saturated carboxylic acids. Another shoulder at 1735 cm^{-1} is characteristic of a saturated aldehyde. However, the major contribution to the carbonyl band is due to the presence of saturated ketones. The broad C=O stretching band is also present in the infrared spectrum of photo-oxidized PE samples, which also show additional bands at 990 and 910 cm^{-1}. The latter bands are characteristic of vinyl groups and their presence shows that chain-terminating unsaturated groups are being formed, most likely as a result of chain scission.

DQ 6.3

Photo- and thermal-degradation of poly(ethylene terephthalate) (PET) both lead to the appearance of the carboxylic acid group, C$_6$H$_5$COOH. The structural repeat unit of PET is illustrated below in Figure 6.19. How may the infrared spectra of PET be used to examine the extent of degradation of samples of this polymer?

Answer

The significant difference between the PET structure and that of the carboxylic acid group degradation product is the presence of the O–H bond in the latter. As more of the degradation product is produced, the intensity of an O–H stretching band in the region of 3300 cm^{-1} will increase.

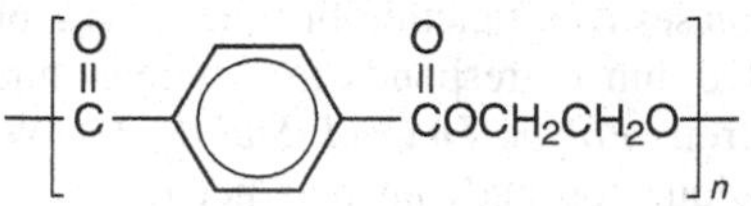

Figure 6.19 Structure of poly(ethylene terephthalate) (cf. DQ 6.3).

Thermogravimetric analysis–infrared (TGA–IR) spectroscopy is a suitable approach to the study of polymer degradation as it allows the degradation pathways to be monitored [23, 24]. A good example of its application is to the study of the degradation of poly(methyl methacrylate) (PMMA) [22]. This polymer is known to degrade via end-chain scission and produce largely monomer as a degradation product. TGA-IR spectroscopy is useful for studying PMMA in the presence of stabilizers such as transition metal halides. When PMMA is mixed in a 1:1 ratio with $FeCl_3$, there is a mass loss of 52% in the 50–350°C

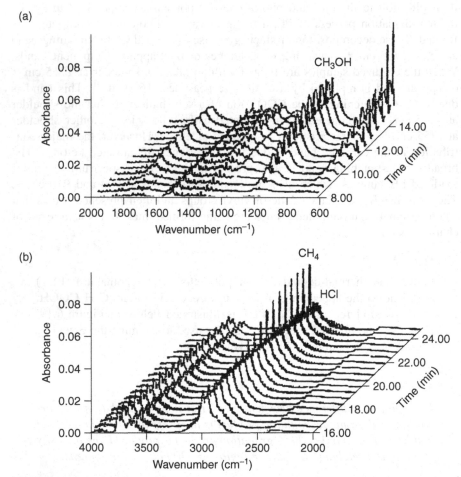

Figure 6.20 Evolution of gases as a function of time from a blend of PMMA and Fe(III) chloride, where a time of 8 min corresponds to a temperature of 160°C (ramp rate of 20°C min^{-1}). Reprinted from *Polym. Degrad. Stabil.*, **66**, Wilkie, C. A., 'TGA/FTIR: An extremely useful technique for studying polymer degradation', 301–306, Copyright (1999), with permission from Elsevier.

temperature range and the evolved gases are H_2O, CH_3OH, MMA and CH_3Cl. In the 350–684°C temperature range, there is a 10% mass loss and CH_3OH, MMA, CH_3Cl, HCl, CH_4 and CO are the evolved gases – there is a residue of 38%. The display of spectra at a number of temperatures in a stacked plot is a good way to follow the progress of the reaction. Figure 6.20 shows a stacked plot for the gases that evolve in the temperature range between 260 and 684°C. This plot illustrates the evolution of gases with time. The evolution of the monomer, signified by the C=O stretching band at 1735 cm^{-1}, begins at about 300°C, while methanol begins to evolve at approximately the same temperature. The evolution of HCl is not observed until the temperature is above 500°C.

Summary

This chapter has demonstrated how infrared spectroscopy may be applied in the study of polymeric systems. First, examples of how the technique may be used to identify and characterize simple polymers, copolymers and blends were provided. Approaches to the quantitative analysis of copolymers and additives or contaminants in polymers were also given. Examples of how polymerization processes may be studied by using infrared spectroscopy were provided. The structural properties of polymers, such as tacticity, branching, crystallinity, hydrogen bonding and orientation, may be readily investigated by using infrared techniques and these were reviewed in this chapter. Also covered were the investigation of the surface properties of polymers and the monitoring of the degradation processes in polymers.

References

1. Bower, D. I. and Maddams, W. F., *The Vibrational Spectroscopy of Polymers*, Cambridge University Press, Cambridge, UK, 1989.
2. Siesler, H. W. and Holland-Moritz, K., *Infrared and Raman Spectroscopy of Polymers*, Marcel Dekker, New York, 1980.
3. Chalmers, J. M., 'Infrared Spectroscopy in Analysis of Polymers and Rubbers', in *Encyclopedia of Analytical Chemistry*, Vol. 9, Meyers, R. A. (Ed.), Wiley, Chichester, UK, 2000, pp. 7702–7759.
4. Chalmers, J. M., Hannah, R. W. and Mayo, D. W., 'Spectra–Structure Correlations: Polymer Spectra', in *Handbook of Vibrational Spectroscopy*, Vol. 3, Chalmers, J. M. and Griffiths, P. R. (Eds), Wiley, Chichester, UK, 2002, pp. 1893–1918.
5. Boerio, F. J., 'Measurements of the Chemical Characteristics of Polymers and Rubbers by Vibrational Spectroscopy', in *Handbook of Vibrational Spectroscopy*, Vol. 4, Chalmers, J. M. and Griffiths, P. R. (Eds), Wiley, Chichester, UK, 2002, pp. 2419–2536.
6. Koenig, J. L., *Spectroscopy of Polymers*, 2nd Edn, Elsevier, Amsterdam, The Netherlands, 1999.
7. Stuart, B. H., *Polymer Analysis*, AnTS Series, Wiley, Chichester, UK, 2002.
8. Chalmers, J. M. and Everall, N. J., 'Qualitative and Quantitative Analysis of Polymers and Rubbers by Vibrational Spectroscopy', in *Handbook of Vibrational Spectroscopy*, Vol. 4, Chalmers, J. M. and Griffiths, P. R. (Eds), Wiley, Chichester, UK, 2002, pp. 2389–2418.

9. Painter, P. C. and Coleman, M. M., *Fundamentals of Polymer Science: An Introductory Text*, Technomic Publishing, Lancaster, PA, USA, 1997.
10. Garton, A., *Infrared Spectroscopy of Polymer Blends, Composites and Surfaces*, Carl Hanser Verlag, Munich, Germany, 1992.
11. Moskala, E. J., Varnell, D. F. and Coleman, M. M., *Polymer*, **26**, 228–234 (1985).
12. Grieve, M. C., *Sci. Justice*, **35**, 179–190 (1995).
13. Bartick, E. G., 'Applications of Vibrational Spectroscopy in Criminal Forensic Science', in *Handbook of Vibrational Spectroscopy*, Vol. 4, Chalmers, J. M. and Griffiths, P. R. (Eds), Wiley, Chichester, UK, 2002, pp. 2993–3004.
14. Tungol, M. W., Bartick, E. G. and Montaser, A., 'Forensic Examination of Synthetic Textile Fibres', in *Practical Guide to Infrared Microspectroscopy*, Humeck, H. (Ed.), Marcel Dekker, New York, 1995, pp. 245–286.
15. Kirkbride, K. P. and Tungol, M. W., 'Infrared Microscopy of Fibres', in *Forensic Examination of Fibres*, Robertson, J. and Grieve, M. (Eds), Taylor and Francis, Philadelphia, PA, USA, 1999, pp. 179–222.
16. Sandler, S. R., Karo, W., Bonesteel, J. and Pearce, E. M., *Polymer Synthesis and Characterization: A Laboratory Manual*, Academic Press, San Diego, CA, USA, 1998.
17. Schroder, E., Muller, G. and Arndt, K. F., *Polymer Characterization*, Carl Hanser Verlag, Munich, Germany, 1989.
18. Mallapragada, S. K. and Narasimham, B., 'Infrared Spectroscopy in Polymer Crystallinity', in *Encyclopedia of Analytical Chemistry*, Vol. 9, Meyers, R. A. (Ed.), Wiley, Chichester, UK, 2000, pp. 7644–7658.
19. Merrill, R. A. and Bartick, E. G., *J. Forensic Sci.*, **45**, 93–98 (2000).
20. Bartick, E. G., Tungol, M. W. and Reffner, J. A., *Anal. Chim. Acta*, **288**, 35–42 (1994).
21. Maynard, P., *J. Forensic Sci.*, **46**, 280–286 (2001).
22. Gardette, J. L., 'Infrared Spectroscopy in the Study of the Weathering and Degradation of Polymers', in *Handbook of Vibrational Spectroscopy*, Vol. 4, Chalmers, J. M. and Griffiths, P. R. (Eds), Wiley, Chichester, UK, 2002, pp. 2514–2522.
23. Wilkie, C. A., *Polym. Degrad. Stabil.*, **66**, 301–306 (1999).
24. Haines, P. J. (Ed.), *Principles of Thermal Analysis and Calorimetry*, The Royal Society of Chemistry, Cambridge, UK, 2002.

Chapter 7
Biological Applications

Learning Objectives

- To choose appropriate sampling techniques for a range of biological applications in infrared spectroscopy.
- To recognize the major infrared bands due to lipids and understand how these may be used to characterize lipid conformation.
- To recognize the characteristic infrared bands associated with proteins and peptides.
- To estimate the secondary structure of proteins by applying quantitative analysis of the infrared spectra of such molecules.
- To recognize the major infrared bands of nucleic acids.
- To use changes in the infrared spectra of animal tissues to characterize disease.
- To recognize how infrared methods may be used to differentiate microbial cells.
- To recognize the common infrared bands observed for plants.
- To understand how near-infrared methods may be used for quantitative analysis in clinical chemistry.

7.1 Introduction

Infrared spectroscopy has proved to be a powerful tool for the study of biological molecules and the application of this technique to biological problems is continually expanding, particularly with the advent of increasingly sophisticated

Infrared Spectroscopy: Fundamentals and Applications   B. Stuart
© 2004 John Wiley & Sons, Ltd   ISBNs: 0-470-85427-8 (HB); 0-470-85428-6 (PB)

sampling techniques such as infrared imaging. Biological systems, including lipids, proteins, peptides, biomembranes, nucleic acids, animal tissues, microbial cells, plants and clinical samples, have all been successfully studied by using infrared spectroscopy [1–11]. This technique has been employed for a number of decades for the characterization of isolated biological molecules, particularly proteins and lipids. However, the last decade has seen a rapid rise in the number of studies of more complex systems, such as diseased tissues. Microscopic techniques, combined with other sophisticated analytical methods, allow for complex samples of micron size to be investigated. This present chapter provides an introduction to the characterization of the main types of biological molecules that may be studied by using infrared techniques.

7.2 Lipids

Infrared spectroscopy can provide valuable structural information about lipids, which are important molecular components of membranes. Many lipids contain phosphorus and are classified as phospholipids, with some examples being shown in Figure 7.1. Lipids are organized in bilayers of about 40–80 Å in thickness, where the polar head group points towards the aqueous phase and the hydrophobic tails point towards the tails of a second layer. The chains can be in an all-*trans*-conformation which is referred to as the *gel phase*, or a *liquid crystalline phase* is obtained when the chain also contains *gauche* C–C groups. The infrared spectra of phospholipids can be divided into the spectral regions that originate from the molecular vibrations of the hydrocarbon tail, the interface region and the head group [12, 13].

The major infrared modes due to phospholipids are summarized in Table 7.1. The hydrocarbon tail gives rises to acyl chain modes. The most intense vibrations in the infrared spectra of lipid systems are the CH_2 stretching vibrations and these give rise to bands in the 3100 to 2800 cm^{-1} region. The CH_2 asymmetric and symmetric

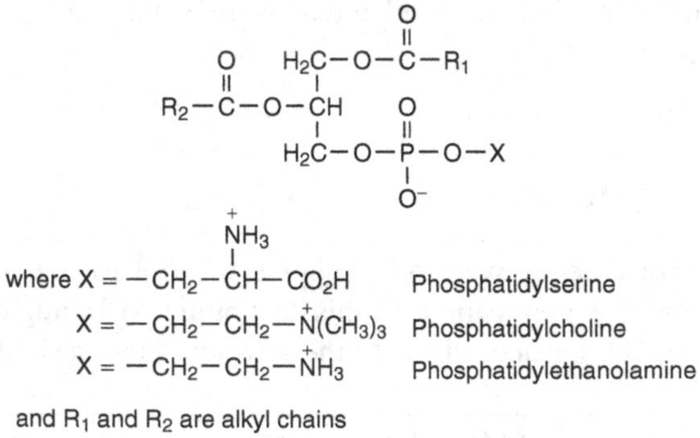

Figure 7.1 Structures of phospholipids.

Table 7.1 Major infrared bands of lipids. From Stuart, B., *Biological Applications of Infrared Spectroscopy*, ACOL Series, Wiley, Chichester, UK, 1997. © University of Greenwich, and reproduced by permission of the University of Greenwich

Wavenumber (cm^{-1})	Assignment
3010	=C–H stretching
2956	CH$_3$ asymmetric stretching
2920	CH$_2$ asymmetric stretching
2870	CH$_3$ symmetric stretching
2850	CH$_2$ symmetric stretching
1730	C=O stretching
1485	(CH$_3$)$_3$N$^+$ asymmetric bending
1473, 1472, 1468, 1463	CH$_2$ scissoring
1460	CH$_3$ asymmetric bending
1405	(CH$_3$)$_3$N$^+$ symmetric bending
1378	CH$_3$ symmetric bending
1400–1200	CH$_2$ wagging band progression
1228	PO$_2^-$ asymmetric stretching
1170	CO–O–C asymmetric stretching
1085	PO$_2^-$ symmetric stretching
1070	CO–O–C symmetric stretching
1047	C–O–P stretching
972	(CH$_3$)$_3$N$^+$ asymmetric stretching
820	P–O asymmetric stretching
730, 720, 718	CH$_2$ rocking

stretching modes, at 2920 and 2851 cm^{-1}, respectively, are generally the strongest bands in the spectra. The wavenumbers of these bands are 'conformation-sensitive' and respond to changes of the *trans/gauche* ratio in the acyl chains. This is also the case for the infrared bands due to the terminal CH$_3$ groups at 2956 cm^{-1} (asymmetric stretching) and 2873 cm^{-1} (symmetric stretching). The =C–H stretching bands due to unsaturated acyl chains are found at 3012 cm^{-1} and the bands due to methylene and methyl groups occur in the 1500–1350 cm^{-1} region. At around 1470 cm^{-1}, there are bands due to CH$_2$ bending and the number and wavenumbers of these bands are dependent on acyl chain packing and conformation. While the asymmetric deformation modes of the CH$_3$ group are obscured by the scissoring bands, the symmetric deformation mode appears at 1378 cm^{-1}.

In certain phospholipid membranes that contain unsaturated acyl chains, the typical lamellar liquid crystalline phase converts to a micellar non-lamellar phase upon heating [14]. Such a thermally induced transition involves a major structural rearrangement. Temperature studies of the infrared spectra of phospholipids provide a sensitive means of studying such transitions in lipids. Figure 7.2 shows the temperature-dependence of the wavenumber of the symmetric CH$_2$ stretching

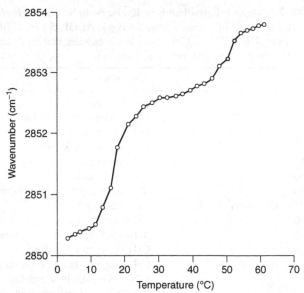

Figure 7.2 Temperature-dependence of the symmetric CH_2 stretching band of phosphatidylethanolamine. From Stuart, B., *Biological Applications of Infrared Spectroscopy*, ACOL Series, Wiley, Chichester, UK, 1997. © University of Greenwich, and reproduced by permission of the University of Greenwich.

band in the spectra of lipid membranes obtained from phosphatidylethanolamine. The increasing wavenumber with temperature indicates an increasing concentration of *gauche*-bands in the acyl chains and this leads to the formation of the non-bilayer phase at higher temperatures. Figure 7.2 shows a wavenumber shift of about 2 cm^{-1} at 18°C and this is associated with the gel-to-liquid crystal phase transition. An additional wavenumber shift of approximately 1 cm^{-1} at 50°C is associated with a transition to the micellar phase. Both of these transitions have been observed to be reversible.

Spectral modes arising from the head group and interfacial region also provide valuable information [15]. Useful infrared bands for studying the interfacial region of lipid assemblies are the ester group vibrations, particularly the C=O stretching bands in the 1750–1700 cm^{-1} region. In diacyl lipids, this region consists of at least two bands originating from the two ester carbonyl groups. A band at 1742 cm^{-1} is assigned to the C=O mode of the first alkyl chain with a *trans*-conformation in the C–C bond adjacent to the ester grouping, while the 1728 cm^{-1} C=O wavenumber of the second alkyl chain suggests the presence of a *gauche*-band in that position. The wavenumber difference observed reflects the structural inequivalence of the chains, with the first alkyl chain initially extending in a direction perpendicular to the second alkyl chain and then developing a *gauche*-bend in order to render the two chains parallel.

SAQ 7.1

The infrared spectrum of phosphatidylserine in deuterium oxide (D$_2$O) was recorded and the deconvolved carbonyl region is shown below in Figure 7.3. What does this band suggest about the conformation of phosphatidylserine?

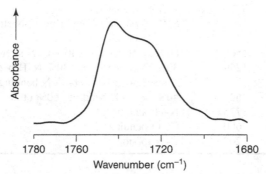

Figure 7.3 The C=O stretching region of phosphatidylserine in D$_2$O (cf. SAQ 7.1). From Stuart, B., *Biological Applications of Infrared Spectroscopy*, ACOL Series, Wiley, Chichester, UK, 1997. © University of Greenwich, and reproduced by permission of the University of Greenwich.

Quantitative infrared analysis can be carried out on blood serum to determine the relative amounts of lipid present [16]. Triglycerides, phospholipids and cholesteryl esters are the classes of lipid that occur in blood serum and such compounds occur naturally in concentrations that make infrared analysis attractive. These classes of compounds can be characterized by their carbonyl bands: the peak maxima appear at 1742 cm^{-1} for the triglycerides, at 1737 cm^{-1} for the phospholipids and at 1723 cm^{-1} for the cholesteryl esters. Although the carbonyl peaks are heavily overlapped, a least-squares method may be used to separate the components. The concentrations of these lipid components are usually in the range 0.03–0.3% in human blood, and standard solutions can be prepared in chloroform.

7.3 Proteins and Peptides

The infrared spectra of proteins exhibit absorption bands associated with their characteristic amide group [17–19]. In-plane modes are due to C=O stretching, C–N stretching, N–H stretching and O–C–N bending, while an out-of-plane mode is due to C–N torsion. The characteristic bands of the amide groups of protein chains are similar to the absorption bands exhibited by secondary amides in general, and are labelled as amide bands. There are nine such bands, called amide A, amide B and amides I–VII, in order of decreasing wavenumber and

Table 7.2 Characteristic infrared amide bands of proteins. From Stuart, B., *Biological Applications of Infrared Spectroscopy*, ACOL Series, Wiley, Chichester, UK, 1997. © University of Greenwich, and reproduced by permission of the University of Greenwich

Designation	Wavenumber (cm^{-1})	Assignment
A	3300 ⎤	N–H stretching in resonance with overtone
B	3110 ⎦	(2 × amide II)
I	1653	80% C=O stretching; 10% C–N stretching; 10% N–H bending
II	1567	60% N–H bending; 40% C–N stretching
III	1299	30% C–N stretching; 30% N–H bending; 10% C=O stretching; 10% O=C–N bending; 20% other
IV	627	40% O=C–N bending; 60% other
V	725	N–H bending
VI	600	C=O bending
VII	200	C–N torsion

these bands are summarized in Table 7.2. Some of the bands are more useful for conformation studies than others and the amide I and amide II bands have been the most frequently used.

The amide II band represents mainly (60%) N–H bending, with some C–N stretching (40%) and it is possible to split the amide II band into components depending on the secondary structure of the protein. The position of the amide II band is sensitive to deuteration, shifting from around 1550 cm^{-1} to a wavenumber of 1450 cm^{-1}. The amide II band of the deuterated protein overlaps with the H–O–D bending vibration, so making it difficult to obtain information about the conformation of this band. However, the remainder of the amide II band at 1550 cm^{-1} may provide information about the accessibility of solvent to the polypeptide backbone. Hydrophobic environments or tightly ordered structures, such as α-helices or β-sheets, reduce the chance of exchange of the amide N–H proton.

The most useful infrared band for the analysis of the secondary structure of proteins in aqueous media is the amide I band, occurring between approximately 1700 and 1600 cm^{-1} [20]. The amide I band represents 80% of the C=O stretching vibration of the amide group coupled to the in-plane N–H bending and C–N stretching modes. The exact wavenumber of this vibration depends on the nature of hydrogen bonding involving the C=O and N–H groups and this is determined by the particular secondary structure adopted by the protein being considered. Proteins generally contain a variety of domains containing polypeptide fragments in different conformations. As a consequence, the observed amide I band is usually a complex composite, consisting of a number of overlapping component bands representing helices, β-structures, turns and random structures.

The infrared contributions of the side chains of the amino acids which constitute the protein must also be considered [21]. Amino acid side chains exhibit infrared modes that are often useful for investigating the local group in a protein. Fortunately, these contributions have been found to be small in D_2O compared to the contributions made by the amide I band. It is also important to be aware of the location of such modes as they may be confused with amide vibrations. The arginyl residue is the only residue that makes a significant contribution in the $1700-1600$ cm^{-1} region, but even the bands at 1586 and 1608 cm^{-1} due to the arginyl residue are small when compared to the amide I contributions. The characteristic side-chain infrared wavenumbers of amino acids are listed in Table 7.3.

Resolution enhancement of the amide I band allows for the identification of various structures present in a protein or peptide. Derivatives and deconvolution can be used to obtain such information. The best method used for the estimation of protein secondary structure involves 'band-fitting' the amide I band. The parameters required, and the number of component bands and their positions, are obtained from the resolution-enhanced spectra. The fractional areas of the fitted component bands are directly proportional to the relative proportions of structure that they represent. The percentages of helices, β-structures and turns may be estimated by addition of the areas of all of the component bands assigned to each of these structures and expressing the sum as a fraction of the total amide I area. The assumption is made that the intrinsic absorptivities of the amide I bands corresponding to different structures are identical.

Infrared spectroscopy has been used in a large number of studies of proteins in a range of environments [22] and the spectral assignments are based on the fact that the secondary structures of these globular proteins have been very well

Table 7.3 Characteristic infrared bands of amino acid side chains. From Stuart, B., *Biological Applications of Infrared Spectroscopy*, ACOL Series, Wiley, Chichester, UK, 1997. © University of Greenwich, and reproduced by permission of the University of Greenwich

Amino acid	Wavenumber (cm^{-1})	Assignment
Alanine	1465	CH$_2$ bending
Valine	1450	CH$_3$ asymmetric bending
Leucine	1375	CH$_3$ symmetric bending
Serine	1350–1250	O–H bending
Aspartic acid	1720	C=O stretching
Glutamic acid	1560	CO$_2^-$ asymmetric stretching
	1415	CO$_2^-$ symmetric stretching
Asparagine	1650	C=O stretching
Glutamine	1615	NH$_2$ bending
Lysine	1640–1610, 1550–1485	NH$_3^+$ bending
	1160, 1100	NH$_3^+$ rocking
Phenylalanine	1602, 1450, 760, 700	Benzene ring vibrations
Tyrosine	1600, 1450	Benzene ring vibrations
Arginine	1608, 1586	Benzene ring vibrations

Table 7.4 Characteristic amide I band assignments of protein secondary structures. From Stuart, B., *Biological Applications of Infrared Spectroscopy*, ACOL Series, Wiley, Chichester, UK, 1997. © University of Greenwich, and reproduced by permission of the University of Greenwich

Wavenumber (cm⁻¹)	Assignment
1695–1670	Intermolecular β-structure
1690–1680	Intramolecular β-structure
1666–1659	'3-turn' helix
1657–1648	α-helix
1645–1640	Random coil
1640–1630	Intramolecular β-structure
1625–1610	Intermolecular β-structure

characterized by X-ray crystallography. The characteristic wavenumbers of the secondary structures of proteins have also been estimated from normal coordinate calculations on model peptides and proteins of known structure [20]. These wavenumbers are summarized in Table 7.4. Such assignments are by no means exact and some bands may appear outside of these ranges for some proteins, particularly when solvent interactions are considered.

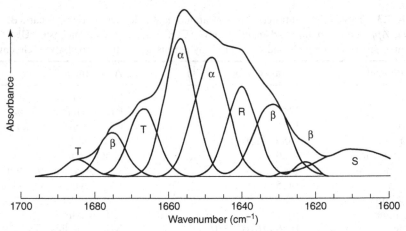

Figure 7.4 Curve-fitted amide I band of lysozyme in D_2O: T, turns and bends; β, β-structure; α, α-helix; R, random coil configuration; S, amino acid side-chain vibrations. From Stuart, B., *Biological Applications of Infrared Spectroscopy*, ACOL Series, Wiley, Chichester, UK, 1997. © University of Greenwich, and reproduced by permission of the University of Greenwich.

Table 7.5 Analysis of the amide I band of
lysozyme in D_2O. From Stuart, B., *Biological Applications of Infrared Spectroscopy*, ACOL Series, Wiley, Chichester,
UK, 1997. © University of Greenwich,
and reproduced by permission of the University of Greenwich

Wavenumber (cm^{-1})	Relative area (%)
1623	1
1632	15
1640	16
1648	24
1657	24
1667	11
1675	7
1684	2
1693	<1

An example of the quantitative approach to protein analysis is illustrated by
Figure 7.4, which shows the amide I band of the enzyme lysozyme in D_2O. The
amide I band of this protein shows nine component bands and the relative areas
of these components are given in Table 7.5. The components may be assigned
to the various types of secondary structures. The bands at 1623 and 1632 cm^{-1}
are characteristic of a β-structure, as is the band at 1675 cm^{-1}. The bands at
1667, 1684 and 1693 cm^{-1} occur at wavenumbers characteristic of turns and
bends, while the 1640 cm^{-1} band may be assigned to a random coil. The band
at 1610 cm^{-1} is due to the arginyl side chain. The two remaining bands at 1648
and 1657 cm^{-1} are due to the presence of an α-helix. Usually in protein infrared
spectra, only one component due to an α-helix is observed. However, X-ray data
indicate the presence of two types of helix in lysozyme, i.e. an α-helix and a
'3-turn' helix. These different helices vibrate at different frequencies and the
1657 cm^{-1} band may be assigned to the α-helix, while the 1648 cm^{-1} band is
due to the '3-turn' helix. These assignments give a quantitative estimate of 48%
helix, 23% β-structure, 13% turns and 16% random coils in lysozyme.

SAQ 7.2

The amide I band of the protein ribonuclease S in D_2O was deconvolved and curve-
fitted and the result is shown below in Figure 7.5. Computer analysis was used to
calculate the relative areas of the components bands illustrated in this spectrum
and the results are given below in Table 7.6. Use these data to approximate the
relative amount of the secondary structures present in ribonuclease S in D_2O.

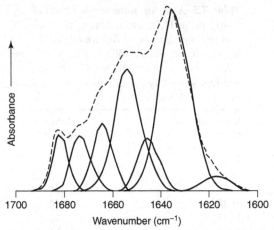

Figure 7.5 Curve-fitted amide I band of ribonuclease S in D_2O (cf. SAQ 7.2). From Stuart, B., *Biological Applications of Infrared Spectroscopy*, ACOL Series, Wiley, Chichester, UK, 1997. © University of Greenwich, and reproduced by permission of the University of Greenwich.

Table 7.6 Analysis of the amide I band of ribonuclease S in D_2O (cf. SAQ 7.2). From Stuart, B., *Biological Applications of Infrared Spectroscopy*, ACOL Series, Wiley, Chichester, UK, 1997. © University of Greenwich, and reproduced by permission of the University of Greenwich

Wavenumber (cm^{-1})	Relative area (%)
1633	43
1645	8
1653	25
1663	10
1672	8
1681	6

Fourier-transform infrared (FTIR) spectroscopy is particularly useful for probing the structures of membrane proteins [3, 23]. This technique can be used to study the secondary structures of proteins, both in their native environment as well as after reconstitution into model membranes. Myelin basic protein (MBP) is a major protein of the nervous system and has been studied by using FTIR spectroscopy in both aqueous solution and after reconstitution in myelin lipids [24]. The amide I band of MBP in D_2O solution (deconvolved and curve-fitted) is

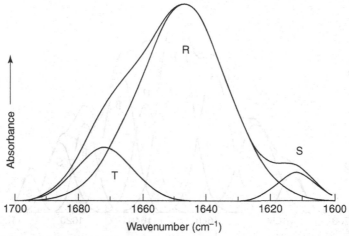

Figure 7.6 Curve-fitted amide I band of myelin basic protein in D_2O: T, turns and bends; R, random coil configuration; S, amino acid side-chain vibrations. From Stuart, B., *Biological Applications of Infrared Spectroscopy*, ACOL Series, Wiley, Chichester, UK, 1997. © University of Greenwich, and reproduced by permission of the University of Greenwich.

Table 7.7 Analysis of the amide I band of myelin basic protein in D_2O and in reconstituted myelin. From Stuart, B., *Biological Applications of Infrared Spectroscopy*, ACOL Series, Wiley, Chichester, UK, 1997. © University of Greenwich, and reproduced by permission of the University of Greenwich

D_2O		Reconstituted myelin	
Wavenumber (cm^{-1})	Relative area (%)	Wavenumber (cm^{-1})	Relative area (%)
		1624	8
		1634	14
		1642	9
1647	86	1651	22
1671	14	1663	14
		1675	11
		1686	3

illustrated in Figure 7.6. The amide band shows a small peak at 1616 cm^{-1}, which occur at a wavenumber normally attributed to amino acid side-chain contributions. The largest component band observed at 1647 cm^{-1} is assigned to the random coil and accounts for 86% of the total band area. The remaining component band at 1671 cm^{-1} is assigned to turns and bends in the protein. Table 7.7 lists the wavenumber and the fractional area of each component of the amide I

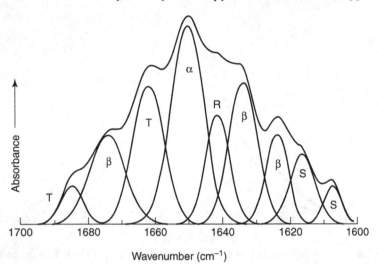

Wavenumber (cm^{-1})

Figure 7.7 Curve-fitted amide I band of myelin basic protein in myelin: T, turns and bends; β, β-structure; α, α-helix; R, random coil configuration; S, amino acid side-chain vibrations. From Stuart, B., *Biological Applications of Infrared Spectroscopy*, ACOL Series, Wiley, Chichester, UK, 1997. © University of Greenwich, and reproduced by permission of the University of Greenwich.

band of MBP in D_2O. The FTIR spectrum of MBP protein after reconstitution in myelin, with all proteins removed from the myelin, shows that there are dramatic changes observed for the amide I band of MBP when the protein is complexed with myelin. The results of deconvolution and band-fitting the amide I band of MBP in myelin are shown in Figure 7.7. The wavenumber and the fractional area of each component of the amide I band of MBP in myelin are also listed in Table 7.7. In Figure 7.7, the two small components at 1606 and 1616 cm^{-1} occur in the wavenumber range attributed to side-chain contributions. The appearance of the band at 1624 cm^{-1} indicates the presence of a water-bonded β-structure. In addition, the component at 1634 cm^{-1} has been assigned to the β-structure, with a high-frequency component due to the β-structure being observed at 1675 cm^{-1}. These bands contribute to 40% of the total band area. This estimation indicates that a notable amount of β-structure is formed when the results are compared to those obtained for MBP in D_2O, where no β-structure was observed. The component at 1651 cm^{-1}, contributing to 28% of the total amide I band area, has been assigned to the α-helix in the protein. Thus, reconstitution of MBP in myelin produces a significant amount of α-helix in the protein in the predominantly lipid environment. Again, none of this type of structure was observed for the protein in an aqueous environment. The small component at 1642 cm^{-1} appears at a wavenumber normally attributed to random coils. The remainder of the components in Figure 7.7 at 1663 and 1686 cm^{-1} may be assigned to turns and bends in the protein. The importance of the conditions in which biological molecules are examined in the infrared needs to be emphasized. The lipid environment more closely mimics the native environment, and thus presumably gives a far better indication of the native structure.

A similar quantitative approach may be applied to peptides. However, their small size must be taken into account when applying such methods, as small peptides are not necessarily capable of producing the same secondary structures observed in proteins. The CiT4 peptide is derived from a nervous system protein and consists of a sequence of 26 amino acids. The amide I band of the CiT4 peptide in D_2O was deconvolved and curve-fitted and the results are illustrated in Figure 7.8. The major component in this figure is observed at 1644 cm^{-1}, contributing to 73% of the total amide I band area. This wavenumber is characteristic of random coils in proteins and peptides. Two smaller component bands appear at 1664 and 1674 cm^{-1}, respectively, and both can be attributed to turns and bends in the peptide structure. The weak component at 1617 cm^{-1} is at a wavenumber associated with arginine side-chain contributions. The solvent environment in which proteins and peptides are recorded affects the secondary structures observed for these molecules. For instance, the solvent trifluoroethanol (TFE) is more polar than water. The FTIR spectrum of the CiT4 peptide in TFE/D_2O was also recorded and is shown in Figure 7.9. There are significant

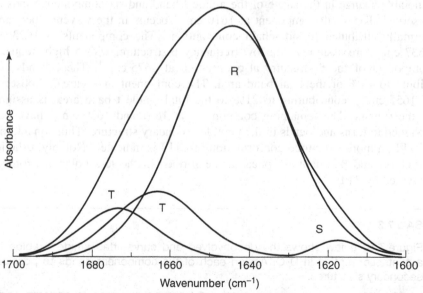

Figure 7.8 Curve-fitted amide I band of CiT4 peptide in D_2O: T, turns and bends; R, random coil configuration; S, amino acid side-chain vibrations. From Stuart, B., *Biological Applications of Infrared Spectroscopy*, ACOL Series, Wiley, Chichester, UK, 1997. © University of Greenwich, and reproduced by permission of the University of Greenwich.

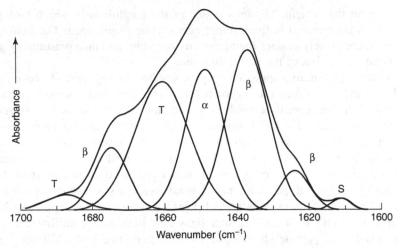

Figure 7.9 Curve-fitted amide I band of CiT4 peptide in TFE/ D_2O: T, turns and bends; β, β-structure; S, amino acid side-chain vibrations. From Stuart, B., *Biological Applications of Infrared Spectroscopy*, ACOL Series, Wiley, Chichester, UK, 1997. © University of Greenwich, and reproduced by permission of the University of Greenwich.

changes observed for the amide I band of CiT4 in the presence of TFE: there is a notable change in the shape of the amide I band and six component bands are observed. The small component at 1610 cm^{-1} occurs in the wavenumber range normally attributed to side-chain contributions. The components at 1623 and 1637 cm^{-1} are assigned to a low-frequency β-structure, with a high-frequency component of the β-structure also observed at 1675 cm^{-1}. These bands contribute to 43% of the total band area. The component in Figure 7.9 observed at 1655 cm^{-1}, contributing to 21% to the total amide I band area, is assigned to the α-helix. The remaining components at 1661 and 1689 cm^{-1} have been assigned to turns and bends in the peptide secondary structure. Thus, on addition of TFE, a more structured conformation of CiT4 is indicated. Notably, α-helical structures and β-structures appear to be formed in the less polar environment provided by TFE.

SAQ 7.3

Figure 7.10 below shows the deconvolved and curve-fitted amide I region of a synthetic peptide in D_2O. Assign each of the component bands to possible secondary structures.

Infrared techniques can also be used to study protein adsorption onto surfaces. For example, a flow-through ATR cell can be used *ex vivo* with flowing blood.

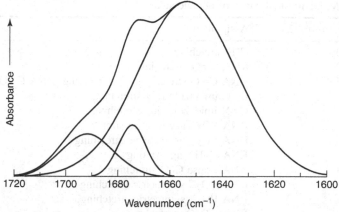

Wavenumber (cm⁻¹)

Figure 7.10 Curve-fitted amide I band of a synthetic peptide in D_2O (cf. SAQ 7.3). From Stuart, B., *Biological Applications of Infrared Spectroscopy*, ACOL Series, Wiley, Chichester, UK, 1997. © University of Greenwich, and reproduced by permission of the University of Greenwich.

Blood–surface interactions are of great importance when medical polymers, such as those used in heart valves and artificial organs, are implanted into the body. When polymers come into contact with blood, complex reactions take place and can result in the formation of a blood clot. Infrared analysis has shown in *ex vivo* studies that during the early stages the proteins albumin and glycoprotein are present, with fibronogen subsequently appearing. As the adsorption process continues, albumin is replaced by other proteins until a blood clot is formed.

7.4 Nucleic Acids

Nucleic acids are vital molecules which carry the genetic code and are responsible for its expression by protein synthesis. The nucleic acids, deoxyribonucleic acid (DNA) and ribonucleic acid (RNA), may be studied by using infrared spectroscopy [25–29]. The spectra of nucleic acids may be divided into the modes due to the constituent base, sugar and phosphate groups. The bases (thymine, adenine, cytosine, guanine and uracil) give rise to purinic and pyrimidinic vibrations in the 1800–1500 cm⁻¹ range and these bands are sensitive markers for base-pairing and base-stacking effects. Bands in the 1500–1250 cm⁻¹ region of nucleic acids are due to the vibrational coupling between a base and a sugar, while in the 1250–1000 cm⁻¹ range sugar–phosphate chain vibrations are observed. These bands provide information about backbone conformations. In the 1000–800 cm⁻¹ region, sugar/sugar–phosphate vibrations are observed. The major infrared modes of nucleic acids are listed in Table 7.8 [30].

Table 7.8 Major infrared bands of nucleic acids [30]

Wavenumber (cm^{-1})	Assignment
2960–2850	CH_2 stretching
1705–1690	RNA C=O stretching
1660–1655	DNA C=O stretching; N–H bending; RNA C=O stretching
1610	C=C imidazole ring stretching
1578	C=N imidazole ring stretching
1244	RNA PO_2^- asymmetric stretching
1230	DNA PO_2^- asymmetric stretching
1218	RNA C–H ring bending
1160, 1120	RNA ribose C–O stretching
1089	DNA PO_2^- symmetric stretching
1084	RNA PO_2^- symmetric stretching
1060, 1050	Ribose C–O stretching
1038	RNA ribose C–O stretching
1015	DNA ribose C–O stretching
	RNA ribose C–O stretching
996	RNA uracil ring stretching; uracil ring bending
970, 916	DNA ribose–phosphate skeletal motions

7.5 Disease Diagnosis

In recent years, infrared spectroscopy has emerged as a powerful tool for characterizing tissue and disease diagnosis [9, 14, 31–38]. The pressure-dependence on parameters such as wavenumber, intensity and bandshape may provide further information about the structural changes associated with malignancy. For example, it is possible to detect cervical cancer arising from a pre-malignant state termed *dysplasia* by using infrared band differences among normal, dysplastic and malignant cervical cells [39]. A study of the pressure-dependence of the wavenumbers of bands indicates that there are extensive changes in the degree of hydrogen bonding of phosphodiester groups of nucleic acids and the C–OH groups of proteins. Pressure studies also indicate changes in the degree of disorder of methylene chains of lipids in the malignant tissue. The infrared spectra of tissue samples with dysplasia demonstrate the same changes to a lesser degree than those observed for the cancer samples. Figure 7.11 shows the pressure-dependence of the CH_2 bending mode wavenumber of the methylene chain of lipids for both normal and malignant tissue. Pressure increases this wavenumber because it induces ordering of the methylene chains in the lipid bilayers, thus increasing inter-chain interactions. In malignant cervical tissue, pressure induces a smaller shift in the wavenumber of this mode when compared with normal tissue. The difference indicates that in cervical cancer the methylene chains of lipids are more disordered than in normal cervical tissue.

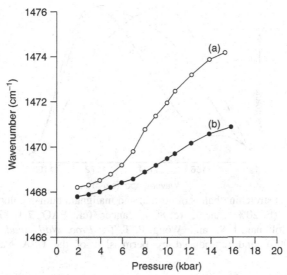

Figure 7.11 Pressure-dependence of the CH$_2$ bending mode wavenumber of the methylene chains of lipids for (a) normal and (b) malignant cervical tissues [39]. From Wong, P. T. T., Wong, R. K., Caputo, T. A., Godwin, T. A. and Rigas, B., *Proc. Natl Acad. Sci., USA*, **88**, 10 988–10 992 (1991), and reproduced by permission of the USA National Academy of Sciences.

SAQ 7.4

FTIR spectroscopy has also been used to study tissue sections of human colorectal cancer [40]. The infrared spectra of three colonic tissue samples with different proportions of normal and malignant components may be obtained and Figure 7.12 below shows the C–O stretching bands of the cell proteins of these samples. What do these spectra suggest about the structural changes in the colonic tissue due to malignancy? (Hint: consider hydrogen bonding.)

Infrared mapping is emerging as a widely used approach for the study of diseased tissues [4, 6, 37, 41]. The identification of diseased regions of tissues is complex and the use of multivariate pattern recognition methods is appropriate. An example of where infrared microscopic mapping may be used to characterize disease is gallstones. Formation of these in the gallbladder is a common disease, particularly in Western countries. The mechanism for the formation of gallstones is still not thoroughly understood and a knowledge of the distribution of the constituents within a 'stone' may provide information about the mechanism of formation. A mixed stone, comprised of concentric bands, has been polished and mapped and some results obtained from this study are illustrated

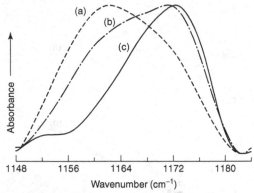

Figure 7.12 C–O stretching bands of normal and malignant human colorectal tissue [40]: (a) 0% cancer; (b) 20% cancer; (c) 80% cancer (cf. SAQ 7.4). From Rigas, B., Morgello, S., Goldman, I. S. and Wong, P. T. T., *Proc. Natl Acad. Sci.*, *USA*, **87**, 8140–8144 (1990), and reproduced by permission of the USA National Academy of Sciences.

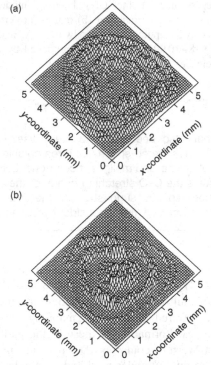

Figure 7.13 Complementary functional group maps of a polished surface of a sectional gallstone: (a) 875 cm^{-1} band of CaCO$_3$; (b) 1050 cm^{-1} band of cholesterol [42]. From Wentrup-Byrne, E., Rintoul, L., Smith, J. L. and Fredericks, P. M., *Appl. Spectrosc.*, **49**, 1028–1036 (1995), and reproduced by permission of the Society for Applied Spectroscopy.

in Figure 7.13 [42]. The concentric bands can be shown to consist of alternating regions of $CaCO_3$ and cholesterol when using infrared spectroscopy. Figure 7.13 shows the complimentary functional group maps of the polished surface of a sectioned gallstone, i.e. (a) a map of the 875 cm^{-1} band of $CaCO_3$ and (b) the 1050 cm^{-1} band of cholesterol.

Kidney disease is relatively common in humans and there are a number of causes of this disease. A knowledge of kidney stone composition and structure is important for establishing its *aetiology*.[†] FTIR spectroscopy has proved successful for the identification of the molecular structure and the crystalline composition of these complex stones [43, 44]. The main components of stones are calcium oxalates, calcium phosphates and purines, although more than 85 mineral or organic compounds have been identified. There are different crystalline forms of these compounds. For example, calcium oxalate forms three crystalline forms in urine: calcium oxalate monohydrate, calcium oxalate dihydrate and calcium oxalate trihydrate. One approach to diagnosis is to examine the form of calcium oxalate in the kidney stone. Examples of the infrared spectra produced for kidney stones of differing calcium oxalate forms are illustrated in Figure 7.14. Here, Figure 7.14(a) shows the spectrum of a stone found to contain 95% calcium oxalate dihydrate, while Figure 7.14(b) represents the spectrum of a stone containing 85% calcium oxalate monohydrate. Quantitative analysis was carried out by using an absorbance ratio of two peaks observed in the spectrum of a binary mixture. The spectrum of the sample in Figure 7.14(a) shows a high proportion of calcium oxalate dihydrate which is linked to a high calcium/oxalate ratio and a high urine calcium content associated with renal hypercalciuria and hyperparathyroidism. In contrast, the predominance of calcium oxalate monohydrate for the spectrum of the sample in Figure 7.14(b) indicates that there is a high oxalate content in the urine, which is linked to primary hyperoxaliuria, inflammatory bowel disease or high intake of oxalate-rich food.

7.6 Microbial Cells

Infrared spectroscopy has proved to be a valuable tool for characterizing and differentiating microbial cells [4, 6, 45–48]. Given the complex nature of such micro-organisms, the spectra can contain a superposition of hundreds of infrared modes. However, the FTIR spectra of bacteria show bands predominantly due to the protein component [6]. Although the cells also contain DNA and RNA structures, carbohydrates and lipids, the various cell and membrane proteins form the major part of the cell mass. Figure 7.15 shows parts of the infrared spectra of *Escherichia coli* and a typical protein, i.e. ribonuclease A, illustrating their similarity. Bands due to carbohydrates and lipids may also be observed in bacterial infrared spectra to a lesser extent.

[†] Aetiology – the medical study of the causation of disease.

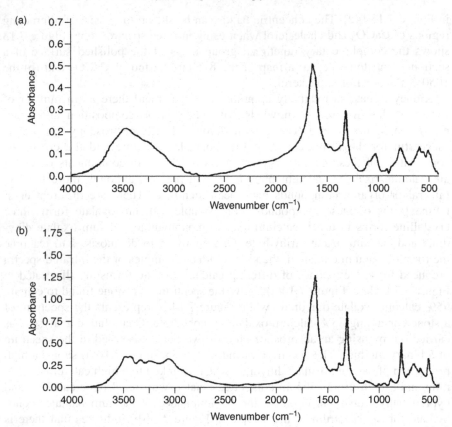

Figure 7.14 Infrared spectra of kidney stones: (a) stone mainly composed of calcium oxalate dihydrate; (b) stone mainly composed of calcium oxalate monohydrate [43]. From Estepa, L. and Daudon, M., *Biospectroscopy*, **3**, 347–369. Copyright © (1997 John Wiley & Sons, Inc). This material is used by permission of John Wiley & Sons, Inc.

FTIR spectroscopy provides a rapid method for identifying micro-organisms responsible for infections. The technique has been used to monitor the biochemical heterogeneity of microcolonies of *E. coli* [49]. By examining the individual spectra and calculating the difference spectra, it is possible to gain a better understanding of the source of the clustering. Figure 7.16 illustrates the infrared spectra obtained from the centre and the edge of an *E. coli* colony. Despite the fact that the spectra appear very similar at first sight, difference spectra do reveal a variation. There are differences in the region near 1230 cm^{-1}, which may be assigned to the phosphate double-bond asymmetric stretching vibration of the phosphodiester, free phosphate and monoester phosphate functional groups. There are also differences observed for the protein amide I region (1670–1620 cm^{-1}), the symmetric stretching vibrations of the COO$^-$ functional groups (1400 cm^{-1}) and

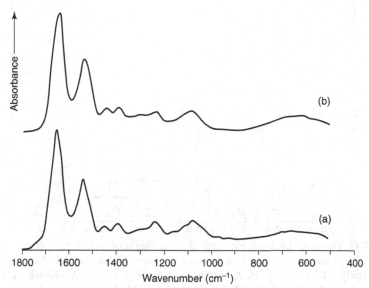

Figure 7.15 Infrared spectra of (a) *Escherichia coli* and (b) ribonuclease [6]. Reprinted from Naumann, D., *Applied Spectroscopy Reviews*, **36**, 239–298 (2001), courtesy of Marcel Dekker, Inc.

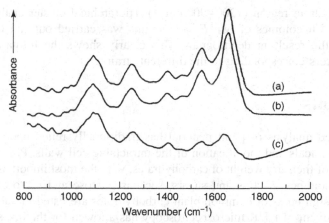

Figure 7.16 Infrared spectra of an *Escherichia coli* colony measured at (a) the centre and (b) the edge of the colony, plus (c) the corresponding difference spectrum [49]. From Choo-Smith, L. P., Maquelin, K., van Vreeswijk, T., Bruining, H. A., Puppels, G. J., Ngo Thi, N. A., Kirschner, C., Naumann, D., Ami, D., Villa, A. M., Orsini, F., Doglia, S. M., Lamfarraj, H., Sockalingum, G. D., Manfait, M., Allouch, P. and Endtz, H. P., *Appl. Environ. Microbiol.*, **67**, 1461–1469 (2001), and reproduced by permission of the American Society for Microbiology.

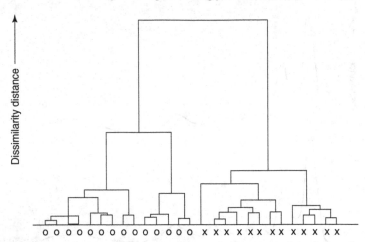

Figure 7.17 Dendrogram from hierarchical clustering analysis of the infrared spectra of 7 h colonies of two *Escherichia coli* strains [49]. From Choo-Smith, L. P., Maquelin, K., van Vreeswijk, T., Bruining, H. A., Puppels, G. J., Ngo Thi, N. A., Kirschner, C., Naumann, D., Ami, D., Villa, A. M., Orsini, F., Doglia, S. M., Lamfarraj, H., Sockalingum, G. D., Manfait, M., Allouch, P. and Endtz, H. P., *Appl. Environ. Microbiol.*, **67**, 1461–1469 (2001), and reproduced by permission of the American Society for Microbiology.

the carbohydrate region (1200–900 cm^{-1}). Hierarchical cluster analysis of the spectra of 7 h colonies of two *E. coli* strains was carried out and Figure 7.17 illustrates the resulting dendrogram. This clearly shows the formation of two major clusters corresponding to the different strains.

7.7 Plants

The infrared analysis of plant material has traditionally relied upon the use of 'harsh' chemicals and modification of the intractable cell walls. Plants comprise up to 80% of their dry weight of carbohydrates, with the most important including cellulose, starches, pectins and sugars, such as glucose and sucrose. The mid-infrared assignments of common plant carbohydrates are listed in Table 7.9 [50, 51]. The advent of FTIR microspectroscopy has allowed for the investigation of localized regions of plant cells which require no modification [50, 52].

An example of where FTIR spectroscopy may be used as a tool for the investigation of plant material is in the examination of the quality of an animal food such as alfalfa. There is a relationship between the quality of alfalfa as a food supplement and the plant age. Infrared microscopy can be used to compare an old plant with a young plant [53]. Figure 7.18 shows the spectra of an old and young leaf (both 100 × 100 μm). At first sight, it is difficult to discern the major differences between these spectra. However, spectral subtraction illustrates some

Table 7.9 Major mid-infrared bands of common plant carbohydrates [50, 51]

Carbohydrate type	Wavenumber (cm^{-1})
α-D-glucose	915, 840
β-D-glucose	915, 900
Methyl α-D-glucopyranoside	900, 845
Methyl β-D-glucopyranoside	850
β-D-fructose	873, 869
β-D-sucrose	910, 869, 850
β-D-cellulose	916, 908
Cellulose	1170–1150, 1050, 1030
Lignin	1590, 1510
Hemicellulose	1732, 1240
Pectin	1680–1600, 1260, 955

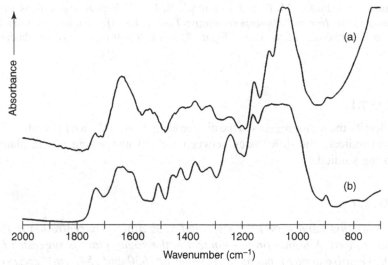

Figure 7.18 FTIR microscopy spectra of leaves from (a) old and (b) young alfalfa plants [53]. Reprinted from Fuller, M. P. and Rosenthal, R. J., 'Biological Applications of FTIR Microscopy', in *Infrared Microspectroscopy: Theory and Applications*, Messerschmidt, R. G. and Harthcock, M. A. (Eds), Figure 6, p. 160 (1988), courtesy of Marcel Dekker, Inc.

important differences and Figure 7.19 shows the difference spectrum obtained by subtracting the spectrum of the young plant from that of the older plant. Positive bands indicate a higher relative concentration in the older leaf, while negative bands indicate a high concentration in the younger plant.

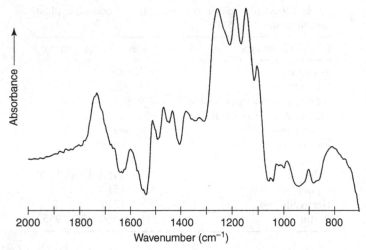

Figure 7.19 Difference spectrum of young and old alfalfa leaves (cf. Figure 7.18) [53]. Reprinted from Fuller, M. P. and Rosenthal, R. J., 'Biological applications of FTIR Microscopy', in *Infrared Microspectroscopy: Theory and Applications*, Messerschmidt, R. G. and Harthcock, M. A. (Eds), Figure 7, p. 161 (1988), courtesy of Marcel Dekker, Inc.

DQ 7.1

Identify the major regions in the difference spectrum shown in Figure 7.19 that indicate the differences between the old and young alfalfa plants being studied.

Answer

A carbonyl band near 1700 cm^{-1} indicates a high concentration in the older plant. A higher protein content in the young plant is suggested by the negative amide I and amide II bands at 1650 and 1545 cm^{-1}, respectively. Thus, Figure 7.19 suggests that there is a significant difference in the cellulosic nature of the two samples, with the older plant having a higher relative cellulose content.

Near-infrared (NIR) spectroscopy is also an excellent tool for examining plant tissue. As in the mid-infrared region, plants produce bands in the NIR region due to the presence of carbohydrates. The latter exhibit overlapped combination and overtone bands of the O–H and C–H groups. Multivariate analysis of the near-infrared spectra of different plant species can be used to differentiate samples [54].

7.8 Clinical Chemistry

Common infrared tests in clinical chemistry include glucose, blood and urine analyses [8, 9, 55–61]. The quantification of glucose in body fluids is an important aspect of clinical analysis, especially in the case of blood glucose measurements for diabetic patients. The use of near-infrared spectroscopy and attenuated total reflectance (ATR) spectroscopy in the mid-infrared region lend themselves to the study of glucose in such aqueous environments, and partial least-squares (PLS) and principal component analysis (PCA) methods have been successfully applied to quantitative analysis of glucose samples [55]. Biological fluids may also be investigated by using NIR or ATR spectroscopy, both as native fluids or as dry films [9]. Figure 7.20 demonstrates the effectiveness of NIR spectroscopy as a technique for the quantitative analysis of native serum samples. This figure shows scatter plots that compare the concentration results obtained by using NIR spectroscopic analysis with the results obtained by using standard-reference clinical methods [9, 62]. In these experiments, the calibration procedure involved using a number of spectra and independent reference analyses, and then to optimize the PLS models for each constituent individually.

DQ 7.2

Figure 7.20 compares serum triglycerides, cholesterol, urea and lactate concentrations obtained by using NIR spectroscopy with 'true' concentrations provided by standard clinical laboratory procedures. Comment on the effectiveness of NIR spectroscopy as a quantitative techniques for each of these serum components.

Answer

The amount of scatter in the plots for triglycerides and urea is minimal, while that for cholesterol is reasonable, and so NIR spectroscopy appears adequate for the analysis of each of these components. However, there is considerable scatter in the plot for lactate, and so NIR spectroscopy is unsuitable for the quantitative analysis of this component in native serum. Lactate appears in quite low concentrations in serum.

Summary

In this chapter, the application of infrared spectroscopy to a range of biological systems was examined. Appropriate sampling techniques and methods of analysis for these systems were discussed. The important infrared bands associated with lipids were introduced and how these relate to the lipid conformation described. The characteristic infrared bands associated with proteins and peptides were outlined. Infrared spectroscopy may be used to estimate the secondary

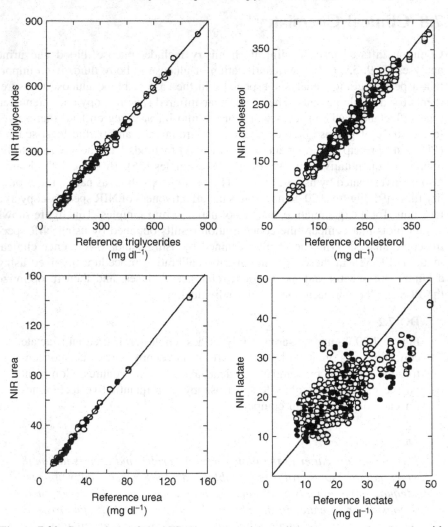

Figure 7.20 Comparison of the NIR (spectroscopic) predicted serum analyte levels with reference analytical results: (o) calibration set; (•) validation set [62]. Reprinted from *Anal. Chim. Acta*, **371**, Hazen, K. H., Arnold, M. A. and Small, G. W., 'Measurement of glucose and other analytes in undiluted human serum with near-infrared transmission spectroscopy', 255–267, Copyright (1998), with permission from Elsevier.

structures of proteins by applying quantitative analysis of the infrared spectra of such molecules and examples were provided. The major infrared bands of nucleic acids were also summarized. Changes in the infrared spectra of animal tissues may be used to characterize disease in such tissues and examples were provided in this chapter. The infrared methods that may be used to differentiate microbial cells were described. Plants may also be investigated by using infrared methods and the common infrared bands observed for plants were summarized. Finally, an understanding of how NIR spectroscopic methods may be used for quantitative analysis in clinical chemistry was presented.

References

1. Clark, R. J. H. and Hester, R. E., *Biomedical Applications of Spectroscopy*, Wiley, Chichester, UK, 1996.
2. Gremlich, H.-U. and Yan, B., *Infrared and Raman Spectroscopy of Biological Materials*, Marcel Dekker, New York, 2000.
3. Jackson, M., Moore, D. J., Mantsch, H. H. and Mendelsohn, R., 'Vibrational Spectroscopy of Membranes', in *Handbook of Vibrational Spectroscopy*, Vol. 5, Chalmers, J. M. and Griffiths, P. R. (Eds), Wiley, Chichester, UK, 2002, pp. 3508–3518.
4. Kalasinsky, V. F., *Appl. Spectrosc. Rev.*, **31**, 193–249 (1996).
5. Mantsch, H. H. and Chapman, D. (Eds), *Infrared Spectroscopy of Biomolecules*, Wiley, New York, 1996.
6. Naumann, D., *Appl. Spectrosc. Rev.*, **36**, 239–298 (2001).
7. Petrich, W., *Appl. Spectrosc. Rev.*, **36**, 181–237 (2001).
8. Shaw, R. A. and Mantsch H. H., 'Vibrational Spectroscopy Applications in Clinical Chemistry', in *Handbook of Vibrational Spectroscopy*, Vol. 5, Chalmers, J. M. and Griffiths, P. R. (Eds), Wiley, Chichester, UK, 2002, pp. 3295–3307.
9. Shaw, R. A. and Mantsch H. H., 'Infrared Spectroscopy in Clinical and Diagnostic Analysis', in *Encyclopedia of Analytical Chemistry*, Vol. 1, Meyers, R. A. (Ed.), Wiley, Chichester, UK, 2000, pp. 83–102.
10. Stuart, B. H., *Biological Applications of Infrared Spectroscopy*, ACOL Series, Wiley, Chichester, UK, 1997.
11. Stuart, B. H., 'Infrared Spectroscopy of Biological Applications', in *Encyclopedia of Analytical Chemistry* Vol. 1, Meyers, R. A. (Ed.), Wiley, Chichester, UK, 2000, pp. 529–559.
12. Watts A. and De Pont J. J. H. H. M. (Eds), *Progress in Protein–Lipid Interactions, 2*, Elsevier, Amsterdam, The Netherlands, 1986.
13. Lewis, R. N. A. H. and McElhaney, R. N., 'Vibrational Spectroscopy of Lipids', in *Handbook of Vibrational Spectroscopy*, Vol. 5, Chalmers, J. M. and Griffiths, P. R. (Eds), Wiley, Chichester, UK, 2002, pp. 3447–3464.
14. Jackson, M. and Mantsch, H. H., *Spectrochim. Acta Rev.*, **15**, 53–69 (1993).
15. Mushayakarara, E. and Levin, I. W., *J. Phys. Chem.*, **86**, 2324–2327 (1982).
16. Bhandara P., Mendelson Y., Stohr E. and Peura, R. A, *Appl. Spectrosc.*, **48**, 271–273 (1994).
17. Bandekar, J., *Biochim. Biophys. Acta*, **1120**, 123–143 (1992).
18. Fabian, H. and Mantele, W., 'Infrared Spectroscopy of Proteins', in *Handbook of Vibrational Spectroscopy*, Vol. 5, Chalmers, J. M. and Griffiths, P. R. (Eds), Wiley, Chichester, UK, 2002, pp. 3399–3425.
19. Fabian, H., 'Fourier Transform Infrared Spectroscopy in Peptide and Protein Analysis', in *Encyclopedia of Analytical Chemistry*, Vol. 7, Meyers, R. A. (Ed.), Wiley, Chichester, UK, 2000, pp. 5779–5803.
20. Arrondo, J. L. R., Muga, A., Castresana, J. and Goni, F. M., *Prog. Biophys. Mol. Biol.*, **59**, 23–56 (1993).
21. Barth, A., *Prog. Biophys. Mol. Biol.*, **74**, 141–173 (2000).
22. Byler, D. M. and Susi H., *Biopolymers*, **25**, 469–487 (1986).
23. Arrondo, J. L. R. and Goni, F. M., *Prog. Biophys. Mol. Biol.*, **72**, 367–405 (1999).

24. Stuart B. H., *Biochem. Mol. Biol. Int.*, **38**, 839–845 (1996).
25. Banyay, M., Sarkar, M. and Graslund, A., *Biophys. Chem.*, **104**, 477–488 (2003).
26. Liquier, J. and Taillandier, E., 'Infrared Spectroscopy of Nucleic Acids', in *Infrared Spectroscopy of Biomolecules*, Mantsch, H. H. and Chapman, D. (Eds), Wiley, New York, 1996, pp. 131–158.
27. Taillandier, E. and Liquier, J., *Methods Enzymol.*, **211**, 307–335 (1992).
28. Taillandier, E. and Liquier, J., 'Vibrational Spectroscopy of Nucleic Acids', in *Handbook of Vibrational Spectroscopy*, Vol. 5, Chalmers, J. M. and Griffiths, P. R. (Eds), Wiley, Chichester, UK, 2002, pp. 3465–3480.
29. Pevsner, A. and Diem, M., *Appl. Spectrosc.*, **55**, 1502–1505 (2001).
30. Benedetti, E., Bramanti, E., Papineschi, F., Rossi, I. and Benedetti E., *Appl. Spectrosc.*, **51**, 792–797 (1997).
31. Jackson, M. and Mantsch, H. H., 'Pathology by Infrared and Raman Spectroscopy', in *Handbook of Vibrational Spectroscopy*, Vol. 5, Chalmers, J. M. and Griffiths, P. R. (Eds), Wiley, Chichester, UK, 2002, pp. 3227–3245.
32. Shaw, R. A. and Mantsch H. H., *J. Mol. Struct.*, **480–481**, 1–13 (1999).
33. Jackson, M. and Mantsch, H. H., '*ex vivo* Tissue Analysis by Infrared Spectroscopy', in *Encyclopedia of Analytical Chemistry*, Vol. 1, Meyers, R. A. (Ed.), Wiley, Chichester, UK, 2000, pp. 131–156.
34. Shultz, C. P., 'Role of Near-Infrared Spectroscopy in Minimally Invasive Medical Diagnosis', in *Handbook of Vibrational Spectroscopy*, Vol. 5, Chalmers, J. M. and Griffiths, P. R. (Eds), Wiley, Chichester, UK, 2002, pp. 3246–3265.
35. Alam, M. K., 'Non-Invasive Diagnosis by Near-Infrared Spectroscopy', in *Handbook of Vibrational Spectroscopy*, Vol. 5, Chalmers, J. M. and Griffiths, P. R. (Eds), Wiley, Chichester, UK, 2002, pp. 3266–3279.
36. Dukor, R. K., 'Vibrational Spectroscopy in the Detection of Cancer', in *Handbook of Vibrational Spectroscopy*, Vol. 5, Chalmers, J. M. and Griffiths, P. R. (Eds), Wiley, Chichester, UK, 2002, pp. 3335–3361.
37. McIntosh, L. M. and Jackson, M., '*ex vivo* Vibrational Spectroscopy Imaging', in *Handbook of Vibrational Spectroscopy*, Vol. 5, Chalmers, J. M. and Griffiths, P. R. (Eds), Wiley, Chichester, UK, 2002, pp. 3376–3387.
38. Sowa, M. G., Leonardi, L. and Matas, A., '*in vivo* Tissue Analysis by Near Infrared Spectroscopy', in *Encyclopedia of Analytical Chemistry*, Vol. 1, Meyers, R. A. (Ed.), Wiley, Chichester, UK, 2000, pp. 251–281.
39. Wong, P. T. T., Wong, R. K., Caputo, T. A. Godwin, T. A. and Rigas, B., *Proc. Natl Acad. Sci., USA*, **88**, 10988–10992 (1991).
40. Rigas, B., Morgello, S., Goldman, I. S. and Wong, P. T. T., *Proc. Natl Acad. Sci., USA*, **87**, 8140–8144 (1990).
41. Attas, M., 'Functional Infrared Imaging for Biomedical Applications', in *Handbook of Vibrational Spectroscopy*, Vol. 5, Chalmers, J. M. and Griffiths, P. R. (Eds), Wiley, Chichester, UK, 2002, pp. 3388–3398.
42. Wentrup-Byrne, E., Rintoul, L., Smith, J. L. and Fredericks, P. M., *Appl. Spectrosc.*, **49**, 1028–1036 (1995).
43. Estepa, L. and Daudon, M., *Biospectroscopy*, **3**, 347–369 (1997).
44. Carmona, P., Bellanato, J. and Escolar, E., *Biospectroscopy*, **3**, 331–346 (1997).
45. Mariey, L., Signolle, J. P., Amiel, C. and Travert, J., *Vibrat. Spectrosc.*, **26**, 151–159 (2001).
46. Maquelin, K., Kirschner, C., Choo-Smith, L. P., van den Braak, N., Endtz, H., Naumann, D. and Puppels, G. J., *J. Microbiol. Methods*, **51**, 255–271 (2002).
47. Schmitt, J. and Flemming H. C., *Int. Biodeter. Biodegrad.*, **41**, 1–11 (1998).

48. Maquelin, K., Choo-Smith, L. P., Kirschner, C., Ngo-Thi, N. A., Naumann, D. and Puppels, G. J., 'Vibrational Spectroscopic Studies of Microorganisms', in *Handbook of Vibrational Spectroscopy*, Vol. 5, Chalmers, J. M. and Griffiths, P. R. (Eds), Wiley, Chichester, UK, 2002, pp. 3308–3334.
49. Choo-Smith, L. P., Maquelin, K., van Vreeswijk, T., Bruining, H. A., Puppels, G. J., Ngo Thi, N. A., Kirschner, C., Naumann, D., Ami, D., Villa, A. M., Orsini, F., Doglia, S. M., Lamfarraj, H., Sockalingum, G. D., Manfait, M., Allouch, P. and Endtz, H. P., *Appl. Environ. Microbiol.*, **67**, 1461–1469 (2001).
50. Mascarenhas, M., Dighton, J. and Arbuckle, G. A., *Appl. Spectrosc.*, **54**, 681–686 (2000).
51. Michell, A. J. and Schimleck, L. R., *Appita J.*, **49**, 23–26 (1996).
52. Stewart, D., *Appl. Spectrosc.*, **50**, 357–365 (1996).
53. Messerschmidt, R. G. and Harthcock, M. A. (Eds), *Infrared Microspectroscopy: Theory and Applications*, Marcel Dekker, New York, 1988.
54. Michell, A. J. and Schimleck, L. R., *Appita J.*, **51**, 127–131 (1998).
55. Heise, H. M., 'Glucose Measurement by Vibrational Spectroscopy', in *Handbook of Vibrational Spectroscopy*, Vol. 5, Chalmers, J. M. and Griffiths, P. R. (Eds), Wiley, Chichester, UK, 2002, pp. 3280–3294.
56. Shaw, R. A., Kotowich, S., Mantsch, H. H. and Leroux, M., *Clin. Biochem.*, **29**, 11–19 (1996).
57. Janatsch, G., Kruse-Jarres, J. D., Marbach, R. and Heise, H. M., *Anal. Chem.*, **61**, 2016–2023 (1993).
58. Shaw, R. A., Kotowich, S., Leroux, M. and Mantsch, H. H., *Clin. Biochem.*, **35**, 634–632 (1998).
59. Ding, Q., Small, G. W. and Arnold, M. A., *Anal. Chem.*, **70**, 4472–4479 (1998).
60. Ng, L. M. and Simmons, R., 'Infrared Spectroscopy in Clinical Chemistry', in *Encyclopedia of Analytical Chemistry*, Vol. 2, Meyers, R. A. (Ed.), Wiley, Chichester, UK, 2000, pp. 1375–1395.
61. Heise, H. M., '*in vivo* Assay of Glucose', in *Encyclopedia of Analytical Chemistry*, Vol. 1, Meyers, R. A. (Ed.), Wiley, Chichester, UK, 2000, pp. 56–83.
62. Hazen, K. H., Arnold, M. A. and Small, G. W., *Anal. Chim. Acta*, **371**, 255–267 (1998).

Chapter 8

Industrial and Environmental Applications

Learning Objectives

- To use infrared spectroscopy to identify and characterize pharmaceutical materials.
- To carry out quantitative analysis of food samples by using infrared spectroscopy.
- To recognize how infrared techniques may be utilized for agricultural applications and in the pulp and paper industries.
- To use infrared spectroscopy to identify the different components of paints.
- To understand how infrared spectroscopy can be used in environmental applications.

8.1 Introduction

The common types of molecules that may be studied using infrared spectroscopy, i.e. organic, inorganic, polymeric and biological, have already been discussed in this book. There is a range of industrial fields that exploit this understanding. This present chapter provides an introduction to a number of important industries and disciplines where infrared spectroscopy has proved to be a useful analytical technique. The disciplines included are the pharmaceutical, food, agricultural, pulp and paper, paint and environmental fields.

Infrared Spectroscopy: Fundamentals and Applications B. Stuart
© 2004 John Wiley & Sons, Ltd   ISBNs: 0-470-85427-8 (HB); 0-470-85428-6 (PB)

8.2 Pharmaceutical Applications

Infrared spectroscopy has been extensively used in both qualitative and quantitative pharmaceutical analysis [1–3]. This technique is important for the evaluation of the raw materials used in production, the active ingredients and the excipients (the inert ingredients in a drug formulation, e.g. lactose powder, hydroxypropyl cellulose capsules, etc.). Although nuclear magnetic resonance spectroscopy and mass spectrometry are widely used in the pharmaceutical industry for the identification of drug substances, infrared spectroscopy can provide valuable additional structural information, such as the presence of certain functional groups.

Figure 8.1 illustrates the diffuse reflectance infrared spectrum of acetylsalicylic acid (aspirin). Such a spectrum may be used to identify the functional groups present in this molecule. The presence of O–H groups in acetylsalicylic acid is indicated by a broad band in the 3400–3300 cm^{-1} region and C–H stretching bands overlap with this band in the 3000–2800 cm^{-1} range. The spectrum shows two strong C=O stretching bands at 1780 and 1750 cm^{-1}, so indicating that the molecule contains carbonyl groups in different environments. The spectrum also shows strong bands at 1150 and 1100 cm^{-1} due to C–O stretching: the 1150 cm^{-1} band is due to the presence of a C–O–H group, while the 1100 cm^{-1} band is due to a C–O–C group in the structure. There is also evidence of a benzene ring, including a series of characteristic bands in the 800–500 cm^{-1} range due to C–H stretching in the aromatic ring. In summary, acetylsalicylic acid contains O–H, C=O, C–O–C, C–O–H and aliphatic and aromatic C–H groups. The acetylsalicylic acid structure is illustrated in Figure 8.2, hence confirming the presence of these functional groups.

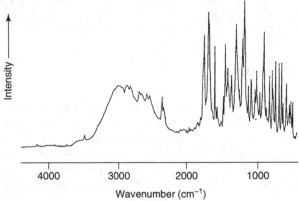

Figure 8.1 Diffuse reflectance infrared spectrum of acetyl salicylic acid. From Stuart, B., *Biological Applications of Infrared Spectroscopy*, ACOL Series, Wiley, Chichester, UK, 1997. © University of Greenwich, and reproduced by permission of the University of Greenwich.

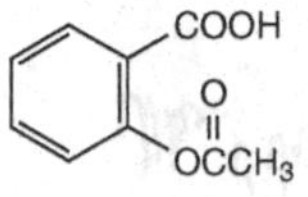

Figure 8.2 Structure of acetylsalicylic acid.

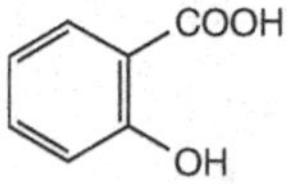

Figure 8.3 Structure of salicylic acid (cf. DQ 8.1).

DQ 8.1

Salicylic acid is a compound used to synthesize aspirin. The structure of this compound is shown below in Figure 8.3. What differences would be expected in the infrared spectra of salicylic acid and acetylsalicylic acid?

Answer

A comparison of the molecular structures of salicylic acid and acetylsalicylic acid informs us that the difference between these two molecules is the presence of an O–H group in the salicylic acid and an acetyl O–COCH₃ group in the acetylsalicylic acid. The crucial differences in the spectra will manifest themselves in the O–H stretching and C=O stretching regions of each spectrum.

SAQ 8.1

The structure of the steroid hormone testosterone is illustrated below in Figure 8.4, with the infrared spectrum of a Nujol mull of testosterone being shown below in Figure 8.5. Assign the major infrared bands of testosterone.

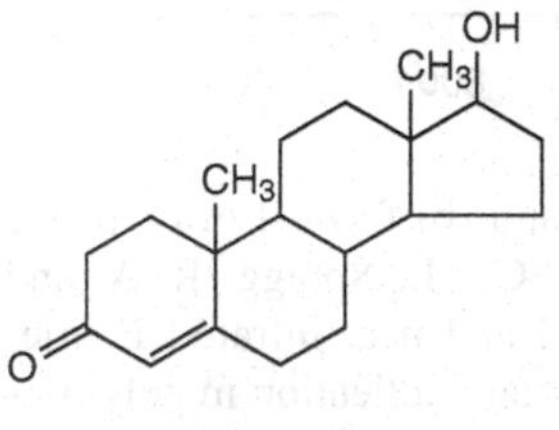

Figure 8.4 Structure of testosterone (cf. SAQ 8.1).

Another important aspect of pharmaceutical analysis is the characterization of the different crystalline forms (polymorphs) of pharmaceutical solids. It is well established that the different forms of a drug can exhibit significantly different physical and chemical properties, and often one form is preferred due to its superior drug activity. An example of how the differences between

170

 Infrared Spectroscopy: Fundamentals and Applications

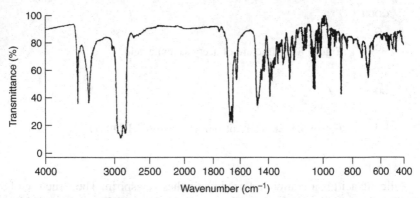

Figure 8.5 Infrared spectrum of testosterone (cf. SAQ 8.1). Reprinted from *Spectrochim. Acta*, **43A**, Fletton, R. A., Harris, R. K., Kenwright, A. M., Lancaster, R. W., Pacher, K. J. and Sheppard, N., 'A comparative Spectroscopic investigation of three pseudopoly-morphs of testosterone using solid-state IR and high resolution solid-state NMR', 1111–1120, Copyright (1987), with permission from Elsevier.

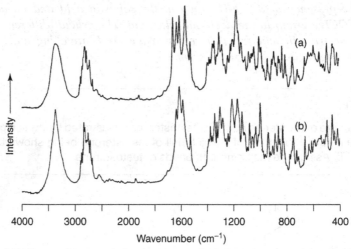

Figure 8.6 Diffuse reflectance infrared spectra of (a) form I and (b) form II of cortisone acetate [4]. Reprinted from *Spectrochim. Acta*, **47A**, Deeley, C. M., Spragg, R. A. and Threlfall, T. L., 'A comparison of Fourier transform infrared and near-infrared Fourier transform Raman spectroscopy for quantitative measurements: an application in polymor-phism', 1217–1223, Copyright (1991), with permission from Elsevier.

polymorphs are observed in their infrared spectra is illustrated in Figure 8.6, in which the diffuse reflectance spectra of two of the polymorphic forms (I and II) of the anti-inflammatory agent, cortisone acetate, are shown [4]. The structure of cortisone acetate is shown in Figure 8.7. Diffuse reflectance is a suitable

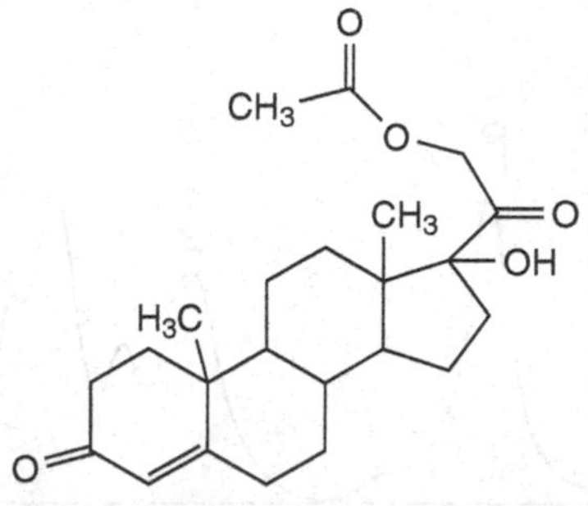

Figure 8.7 Structure of cortisone acetate.

sampling technique because it avoids the structural changes that may occur under high pressures during the formation of the alkali halide discs. Although the spectra shown in Figure 8.6 are complex below 1500 cm^{-1}, in the region above 1500 cm^{-1} it is easier to identify differences in the spectra of the two polymorphs. In the 1800–1500 cm^{-1} region, there are significant differences in band intensities. There is also a notable difference in the appearance of the O–H stretching band in the 3600–3100 cm^{-1} region.

SAQ 8.2

An anti-inflammatory agent, the structure of which is shown below in Figure 8.8, forms two crystalline polymorphs (I and II) that have been differentiated by microscopic techniques and show different properties [5]. The carbonyl regions of the diffuse reflectance spectra of the two polymorphs are shown below in Figure 8.9. Comment on the differences in the carbonyl regions in the spectra of these polymorphs. Suggest a simple method for quantitatively determining the amount of polymorph II in a mixture of the two polymorphs of this drug.

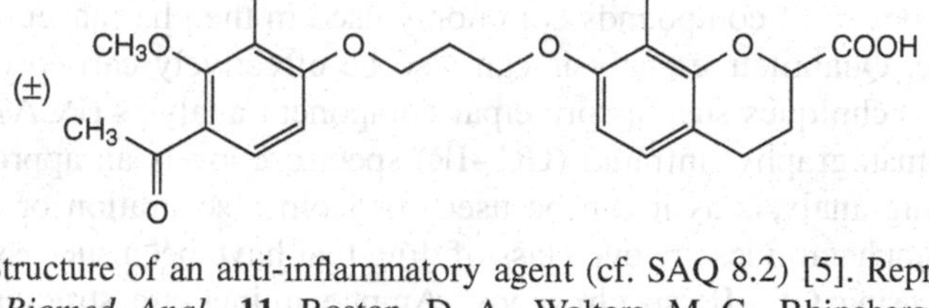

Figure 8.8 Structure of an anti-inflammatory agent (cf. SAQ 8.2) [5]. Reprinted from *J. Pharmaceut. Biomed. Anal.*, **11**, Roston, D. A., Walters, M. C., Rhinebarger, R. R. and Ferro, L. J., 'Characterization of polymorphs of a new anti-inflammatory drug', 293–300, Copyright (1993), with permission from Elsevier.

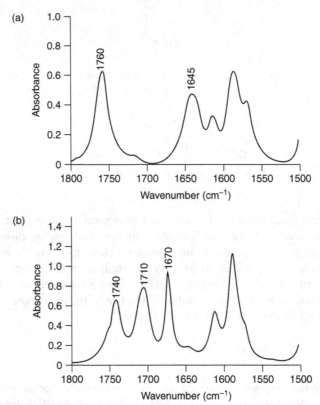

Figure 8.9 Carbonyl regions of the diffuse reflectance infrared spectra of two poly-morphs of an anti-inflammatory drug: (a) polymorph I; (b) polymorph II (cf. SAQ 8.2) [5]. Reprinted from *J. Pharmaceut. Biomed. Anal.*, **11**, Roston, D. A., Walters, M. C., Rhinebarger, R. R. and Ferro, L. J., 'Characterization of polymorphs of a new anti-inflammatory drug', 293–300, Copyright (1993), with permission from Elsevier.

Near-infrared (NIR) spectroscopy lends itself to the pharmaceutical quality control laboratory [6, 7]. The development of fibre optic probes for remote anal-ysis has lead to the expansion of its use in the pharmaceutical industry. Libraries of the NIR spectra of compounds commonly used in the pharmaceutical industry are available. Quantitative analysis can also be effectively carried out by using multivariate techniques such as principal component analysis (PCA).

Gas chromatography–infrared (GC–IR) spectroscopy is an appropriate tech-nique for drug analysis as it can be used for isomer separation or contaminant detection. Amphetamines are one class of drug that have been successfully differ-entiated by using GC–IR spectroscopy. Amphetamines are structurally similar molecules that can be easily mis-identified. Although such similar compounds cannot be differentiated by their mass spectra, there are prominent differences

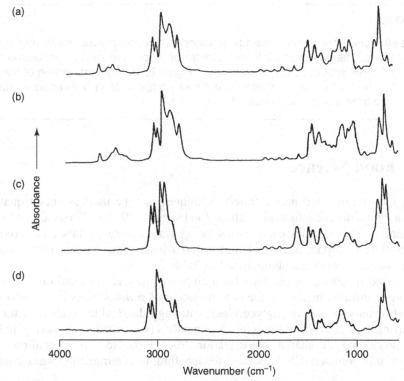

Figure 8.10 Infrared spectra of the components of a clandestine laboratory drug mixture separated by GC–IR spectroscopy: (a) ephedrine; (b) pseudoephedrine; (c) amphetamine; (d) methamphetamine [8]. From Bartick, E. G., 'Applications of Vibrational Spectroscopy in Criminal Forensic Analysis', in *Handbook of Vibrational Spectroscopy*, Vol. 4, Chalmers, J. M. and Griffiths, P. R. (Eds), pp. 2993–3004. Copyright 2002. © John Wiley & Sons Limited. Reproduced with permission.

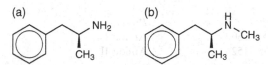

Figure 8.11 Structures of (a) amphetamine and (b) methamphetamine.

in their infrared spectra. Figure 8.10 illustrates the gas-phase FTIR spectra of a clandestine laboratory mixture separated by GC–IR spectroscopy. The spectra of amphetamine and methamphetamine are illustrated and the structures of these two compounds are shown in Figure 8.11. The spectra importantly differ in the region below 1700 cm^{-1} [8]. The N–H bending band near 1600 cm^{-1} is more intense for amphetamine, which is a primary amine.

SAQ 8.3

The stereoisomers, ephedrine and pseudoephedrine, are precursors for metham-phetamine. The gas-phase infrared spectra of these compounds, separated by using GC–IR spectroscopy, are shown in Figure 8.10. From inspection of these spectra, is it possible to differentiate these compounds in a forensic sample obtained from a clandestine laboratory?

8.3 Food Science

Both mid-infrared and near-infrared techniques may be used to obtain qualitative and quantitative information about food samples [9–13]. Foods are complex mixtures, with the main components being water, proteins, fats and carbohydrates. The important analytical mid-infrared bands associated with the major components of foods are summarized in Table 8.1.

Infrared spectroscopy has long been employed to study fats and oils [9]. One important infrared method is the determination of *trans*-isomers in fats and oils (which mainly consist of triglycerides). Although the double bonds in naturally occurring fats and oils show a predominantly *cis*-configuration, during industrial processing substantial isomerization from a *cis*- to a *trans*-configuration occurs. It is commercially important for labelling to determine the *trans*-content

Table 8.1 Mid-infrared bands of major food components

Wavenumber (cm^{-1})	Assignment
	Water
3600–3200	O–H stretching
1650	H–O–H stretching
	Proteins
1700–1600	Amide I
1565–1520	Amide II
	Fats
3000–2800	C–H stretching
1745–1725	C=O stretching
967	C=C–H bending
	Carbohydrates
3000–2800	C–H stretching
1400–800	Coupled stretching and bending

of fats and oils. The *cis*- and *trans*-isomers show distinctive C=C–H out-of-plane bending bands in the mid-infrared spectra. While *cis*-isomers show bands in the 840 to 700 cm^{-1} region, the *trans*-isomer shows a characteristic peak at 966 cm^{-1}. Both isomers show a strong band at 1163 cm^{-1} due to C–O stretching of the ester group, which may be used as a reference band for quantitative analysis.

DQ 8.2

Describe how the percentage *trans*-content of a sample of partially hydrogenated vegetable oil could be estimated by using mid-infrared spectroscopy.

Answer

*A series of oil samples of known **trans**-composition should be prepared. The infrared spectra of each standard should be recorded and the ratio of the absorbances of the 996 and 1163 cm^{-1} bands measured. A plot of this ratio versus the percentage **trans**-content of the samples can then be used as a calibration plot. The same absorbance ratio from the infrared spectrum of the sample of unknown **trans**-content can be measured and the calibration plot used to determine the percentage **trans**-content of the unknown sample.*

Near-infrared (NIR) spectroscopy is widely used in the food industry as a fast routine analytical method for the quantitative measurement of water, proteins, fats and carbohydrates [13]. Although the near-infrared bands are less useful for qualitative analysis of foods because of their broad overlapped appearance, these bands are suitable for quantitative analysis when using chemometric techniques. Figure 8.12 illustrates the appearance of the major food components in the near-infrared, showing the spectrum of a sample of dehydrated tomato soup.

NIR spectroscopy has been used for many years in the dairy industry to examine milk [12, 14]. While milk consists of more than 80% water, the concentrations of the fat and protein components have a significant effect on the near-infrared spectrum. The spectra of milk samples shift to lower wavelengths as the fat and protein concentrations decrease. This is due to changes in the amount of light scattering by the fat globules and protein micelles.

Wine analysis is also feasible with near-infrared spectroscopy [12]. This approach provides a relatively straightforward method for analysing ethanol in wine. The O–H bond in ethanol produces a near-infrared band that is easily distinguished from the O–H band of water. In the 1980s, the wine industry was troubled by several cases of adulteration using diethylene glycol and methanol. Methanol is easily distinguished from ethanol in the near-infrared. Figure 8.13

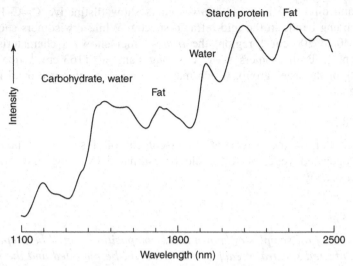

Figure 8.12 Near-infrared spectrum of dehydrated tomato soup. From Davies, A. M. C. and Grant, A., 'Near Infrared Spectroscopy for the Analysis of Specific Molecules in Food', in *Analytical Applications of Spectroscopy*, Creaser, C. M. and Davies, A. M. C. (Eds), pp. 46–51 (1998). Reproduced by permission of The Royal Society of Chemistry.

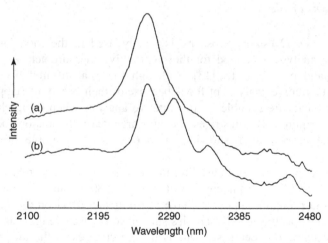

Figure 8.13 Near-infrared spectra of wine samples containing (a) 1% methanol and (b) 1% ethanol. From Davies, A. M. C. and Grant, A., 'Near Infrared Spectroscopy for the Analysis of Specific Molecules in Food', in *Analytical Applications of Spectroscopy*, Creaser, C. M. and Davies, A. M. C. (Eds), pp. 46–51 (1998). Reproduced by permission of The Royal Society of Chemistry.

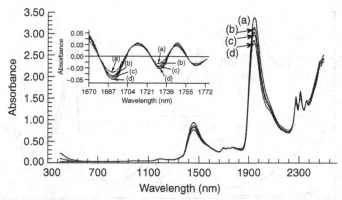

Figure 8.14 Near-infrared spectra of four commercial spirits: (a) Scotch whiskey (40% ethanol); (b) gin (47% ethanol); (c) vodka (50% ethanol); (d) Bourbon (55% ethanol) [12]. From McClure, W. F. and Stanfield, D. L., 'Near-Infrared Spectroscopy of Biomaterials', in *Handbook of Vibrational Spectroscopy*, Vol. 1, Chalmers, J. M. and Griffiths, P. R. (Eds), pp. 212–225. Copyright 2002. © John Wiley & Sons Limited. Reproduced with permission.

illustrates that a 1% addition of methanol to a wine sample is easily distinguished from a 1% addition of ethanol.

Other alcoholic beverages, including beers and spirits, can be analysed by using NIR spectroscopy. Figure 8.14 illustrates the NIR spectra of four commercial spirits, i.e. Scotch whiskey (40% ethanol), gin (47% ethanol), vodka (50% ethanol) and Bourbon (55% ethanol), with the inset showing the corresponding second-derivative bands of the region near 1700 nm (first overtone of C–H stretching). Therefore, despite the fact that it is difficult to visually discern differences between each spectrum in the main figure, the derivatives help to show that there are significant differences between these spectra, which is dependent upon the alcohol concentration.

SAQ 8.4

Figure 8.15 below presents the NIR spectra of four different commercial spirits, namely Tequila, a liqueur, rum and Cognac, along with an inset showing the second derivatives of these spectra in the 1700 nm region. The samples are of different ethanol concentrations, i.e. 17, 22, 34 and 40%, respectively. Which spectrum corresponds to which sample concentration?

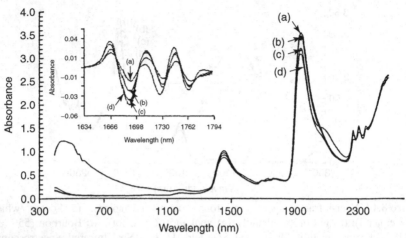

Figure 8.15 Near-infrared spectra of four commercial spirits (cf. SAQ 8.4) [12]. From McClure, W. F. and Stanfield, D. L., 'Near-Infrared Spectroscopy of Biomaterials', in *Handbook of Vibrational Spectroscopy*, Vol. 1, Chalmers, J. M. and Griffiths, P. R. (Eds), pp. 212–225. Copyright 2002. © John Wiley & Sons Limited. Reproduced with permission.

8.4 Agricultural Applications

Commercial grains are commonly analysed by using NIR spectroscopy [12, 15]. The major constituents of grains are water, protein, oil, fibre, minerals and carbohydrates and it is commercially important to quantitate the composition. The NIR spectra of such materials are, thus, dominated by the overtones and combination bands of C–H, N–H, O–H and C=O bonds. Figure 8.16 illustrates the NIR spectra (in reflection mode) of a number of common grains, i.e. sesame seeds, sunflower seeds, wheat, barley, grass and oats.

DQ 8.3

Given that sesame and sunflower seeds are well known to contain a high oil content, comment on the difference observed between the NIR spectra of these seeds and those of the other grains illustrated in Figure 8.16.

Answer

As oils consist of long hydrocarbon chains, the NIR spectra of oils show prominent bands due to the C–H bond. This is evident in the spectra obtained for the sesame and sunflower seeds shown in Figure 8.16. Notable bands due to the first overtone of the C–H stretching band can be seen at 1720 and 1760 nm, while combination bands of C–H stretching and bending appear at 2308 and 2346 nm.

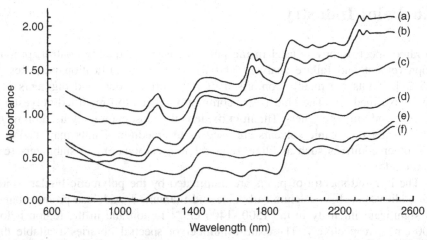

Figure 8.16 Near-infrared reflection spectra of grains: (a) sesame seed; (b) sunflower; (c) wheat; (d) barley; (e) grass; (f) oats [12]. From McClure, W. F. and Stanfield, D. L., 'Near-Infrared Spectroscopy of Biomaterials', in *Handbook of Vibrational Spectroscopy*, Vol. 1, Chalmers, J. M. and Griffiths, P. R. (Eds), pp. 212–225. Copyright 2002. © John Wiley & Sons Limited. Reproduced with permission.

8.5 Pulp and Paper Industries

Infrared spectroscopy plays an important role in quality control in the pulp and paper industries [16, 17]. Paper is comprised of cellulose fibres, inorganic fillers and binders. In the mid-infrared region, additives, such as polymers and calcium carbonate, may be identified in paper. The majority of bands in a paper spectrum are due to cellulose. The latter shows an O–H stretching band near 3330 cm^{-1}, C–H stretching bands in the 3000–2900 cm^{-1} region, C–H bending bands in the range 1500–1300 cm^{-1}, and a C–O ether stretching band at 1030 cm^{-1}.

Although water is an important component of paper, water interference in the infrared spectra is minimized by the used of attenuated total reflectance (ATR) spectroscopy. The diffuse reflectance infrared technique (DRIFT) is also particularly helpful for the examination of paper samples as it may be utilized for coarse samples that are difficult to grind into fine powders. Photoacoustic spectroscopy (PAS) may also be used for paper analysis, as additives such as calcium carbonate exhibit strong photoacoustic spectra. In addition, because of the heterogeneous nature of paper, FTIR microscopy in an effective approach to the analysis of such samples. NIR spectroscopy has emerged as another approach to paper analysis during quality control, particularly because of the ability to use this technique for remote analysis. Furthermore, multivariate calibration methods can be used to determine the composition of components such as lignin, carbohydrates and inorganic additives.

8.6 Paint Industry

Infrared spectroscopy is used in the paint industry for quality control, product improvement and failure analysis, and for forensic identification purposes [8, 18–21]. Paints are mainly comprised of polymeric binders and pigments in a dispersive medium. The binders are commonly alkyds (oil-modified polyesters), acrylics and vinyl polymers. Titanium oxide is the most commonly used pigment, while water or organic solvents are used as the medium. Paints may also contain other additives such as fillers (e.g. calcium carbonate) and stabilizers (e.g. lead oxide).

The infrared spectra of paints are dominated by the polymeric binder bands, but additives such as calcium carbonate and titanium dioxide can produce bands of significant intensity in the $1500–1400$ cm^{-1} region and in the region below 800 cm^{-1}, respectively. There are a number of spectral libraries available that may be used as reference sources for the identification of paints and coatings [22, 23].

The most common infrared sampling techniques used to examine paint samples are attenuated total reflectance and photoacoustic spectroscopies. Liquid paints, dried films and paint chips may all be investigated in this way. Depth profiling can be useful when examining paint, as the surface properties will vary importantly from the bulk properties. In addition, most paint films contain two or more layers with different compositions, and so reflectance techniques are necessary for the characterization of individual layers.

SAQ 8.5

The ATR spectrum of a white water-based acrylic paint is shown below in Figure 8.17. There is a variety of acrylic binder structures possible, but all contain an ester group. Identify the major bands in this spectrum.

Infrared spectroscopy may also be employed to identify pigments in paints, which is of particular importance in the fields of art conservation and forensic analysis [8, 24–26]. Colourants and dyes are commonly inorganic materials and there are collections of the infrared spectra recorded for many colourants [27–29]. Infrared microscopic techniques are clearly of great use when examining paint chips, either for forensic samples or precious fragments obtained from rare artworks. For identification of a pigment component in a paint sample, it is helpful to use, e.g. spectral subtraction or solvent extraction procedures to separate the binder from pigments before analysis. An example of an infrared spectrum of a common dye obtained by using infrared microscopy is shown in Figure 8.18.

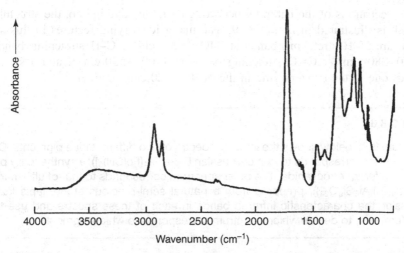

Figure 8.17 Attenuated total reflectance spectrum of a white acrylic paint (cf. SAQ 8.5) [18]. From Carr, C., 'Vibrational Spectroscopy in the Paint Industry', in *Handbook of Vibrational Spectroscopy*, Vol. 4, Chalmers, J. M. and Griffiths, P. R. (Eds), pp. 2935–2951. Copyright 2002. © John Wiley & Sons Limited. Reproduced with permission.

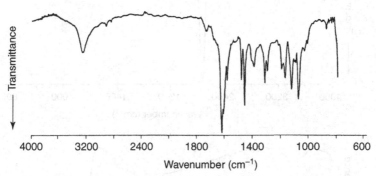

Figure 8.18 Infrared spectrum of indigo [24]. Reproduced from Derrick, M. R., Stulik, D. and Landry, J. M., *Infrared Spectroscopy in Conservation Science*, p. 197, 1999, by permission of the Getty Conservation Institute. © *The J. Paul Getty Trust, [1999]. All rights reserved*.

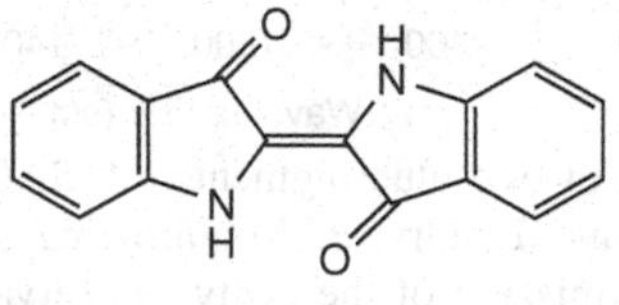

Figure 8.19 Structure of indigo.

The spectrum is of the naturally occurring dark blue dye indigo, the structure of which is illustrated in Figure 8.19. The major identifying features in this spectrum are N–H stretching bands at 3400–3200 cm^{-1}, C–H stretching bands at 3100–2800 cm^{-1}, C=O stretching bands at 1700–1550 cm^{-1}, and a series of bands due to the benzene ring in the 1620–1420 cm^{-1} region.

SAQ 8.6

Figure 8.20 below shows the infrared spectra of two different blue pigments. One spectrum corresponds to that of Prussian Blue ($Fe_4[Fe(CN)_6]$), a synthetically produced ferric ferrocyanide. The other spectrum corresponds to that of ultramarine ($Na_8(S_2)(Al_6Si_6O_{24})$), prepared from a natural semi-precious stone, lapis lazuli. Assign the characteristic infrared bands in each of these spectra and use this information to decide which spectrum corresponds to which pigment.

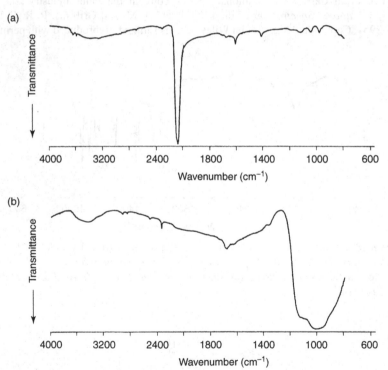

Figure 8.20 Infrared spectra of two blue pigments (cf. SAQ 8.6) [24]. Reproduced from Derrick, M. R., Stulik, D. and Landry, J. M., *Infrared Spectroscopy in Conservation Science*, p. 198, 1999, by permission of the Getty Conservation Institute. © *The J. Paul Getty Trust, [1999]. All rights reserved.*

8.7 Environmental Applications

Infrared spectroscopy has been applied to a broad range of environmental sampling problems, including air, water and soil analysis [30]. The development of remote sensing infrared techniques has been particularly useful in this field and simple practical infrared techniques have been developed for measuring trace gases and atmospheric compositions [30–34]. Common applications include industrial gas emissions, emissions from fires, and astronomical applications. Determination of the compositions of atmospheric gases is important for an understanding of global climate changes and infrared spectroscopy can be used for measuring the most abundant and important greenhouse gases, i.e. CO_2, CH_4, N_2O and CO [32].

Routine monitoring of the vapour levels of toxic chemicals in the work environment has led to the development of infrared techniques for the automatic quantitative analysis of noxious chemicals in air [30, 32]. The technique is based on long-pathlength cells with a pump to circulate the air sample through the instrument. Such instruments are dedicated to a single contaminant in air, employing a filter system for wavelength selection, and these are commonly used for SO_2, HCN, phosgene, NH_3, H_2S, formaldehyde and HCl. Other instruments are *tunable* for a particular analytical band and usually give a direct read-out in parts per million (ppm) or volume percentage (vol%) of the pollutant, while the pathlength of the cell can also be varied. Typical applications are the monitoring of formaldehyde in plastic and resin manufacture, anaesthetics in operating theatres, degreasing solvents in a wide variety of industries, and carbon monoxide in garages.

SAQ 8.7

Figure 8.21 below illustrates an infrared spectrum of cigarette smoke [35], obtained using a 14 cm gas cell with KBr windows, at a resolution of 1 cm^{-1}. Given that the main components of cigarette smoke are water, carbon monoxide, carbon dioxide, hydrogen cyanide and methane, assign the major peaks shown in this figure. Are there any bands in the spectrum that cannot be assigned to these compounds?

Infrared spectroscopy has proved to be a very effective tool for the investigation of solid pollutants. For example, contaminants, such as pesticides, organic and inorganic sprays, or dusts on plants, may be examined by using this technique. GC–IR spectroscopy has been used to investigate the degradation products of 'Mirex' (dodecachloropentacyclodecane), a pesticide used in the control of fire ants [36]. One of the requirements for pesticides is that they rapidly decompose in the environment, and GC–IR spectroscopy can be used to identify 'Mirex' derivatives. This technique has an advantage over GC–MS, for instance,

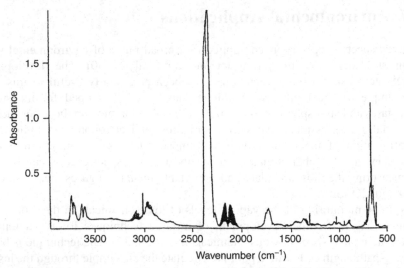

Figure 8.21 Infrared spectra of cigarette smoke (cf. SAQ 8.7) [35]. Used with permission from the *Journal of Chemical Education*, **78**, No. 12, 2001, pp. 1665–1666; Copyright ©2001, Division of Chemical Education, Inc.

because it can more readily differentiate the derivatives produced as a result of degradation. During the decomposition of 'Mirex', some of the chlorine atoms are substituted with hydrogen. The 5,10-dichloro derivative has *anti-* and *syn-* isomers, and these are difficult to separate by using GC–MS. However, the infrared spectra of the two isomers show bands at 1100 and 1120 cm^{-1}, respectively. There is some overlap of these bands, but the pure spectrum of each isomer can be obtained from the difference spectra.

Summary

This chapter has provided an introduction to a number of industrial (and related) fields which utilize infrared spectroscopy as an analytical technique. This method is widely used in the pharmaceutical industry for the qualitative and quantitative analysis of active and non-active ingredients. The food industry uses information from the mid- and near-infrared regions to carry out qualitative and quantitative analysis. Agricultural applications, such as the evaluation of grain, and the pulp and paper industries, were introduced and near-infrared spectroscopy was demonstrated as an important approach in these fields. Paints are variable mixtures and infrared spectroscopy provides an effective technique for the identification of the components of paints. Examples of environmental applications of infrared spectroscopy, including gases and pollutants, were also discussed.

References

1. Clark, D., 'The Analysis of Pharmaceutical Substances and Formulated Products by Vibrational Spectroscopy', in *Handbook of Vibrational Spectroscopy*, Vol. 5, Chalmers, J. M. and Griffiths, P. R. (Eds), Wiley, Chichester, UK, 2002, pp. 3574–3589.
2. Bugay, D. E., *Adv. Drug Delivery Rev.*, **48**, 43–65 (2001).
3. Aldrich, D. S. and Smith, M. A., *Appl. Spectrosc. Rev.*, **34**, 275–327 (1999).
4. Deeley, C. M., Spragg, R. A. and Threlfall, T. L., *Spectrochim. Acta*, **47A**, 1217–1223 (1999).
5. Roston, D. A., Walters, M. C., Rhinebarger, R. R. and Ferro, L. J., *J. Pharmaceut. Biomed. Anal.*, **11**, 293–300 (1993).
6. Blanco, M., Coello, J., Iturriaga, H., Maspoch, S. and de la Pezuela, C., *Analyst*, **123**, 135R–150R (1998).
7. Broad, N., Graham, P., Hailey, P., Hardy, A., Holland, S., Hughes, S., Lee, D., Prebbe, K., Salton, N. and Warren, P., 'Guidelines for the Development and Validation of Near-Infrared Spectroscopic Methods in the Pharmaceutical Industry', in *Handbook of Vibrational Spectroscopy*, Vol. 5, Chalmers, J. M. and Griffiths, P. R. (Eds), Wiley, Chichester, UK, 2002, pp. 3590–3610.
8. Bartick, E. G., 'Applications of Vibrational Spectroscopy in Criminal Forensic Analysis', in *Handbook of Vibrational Spectroscopy*, Vol. 4, Chalmers, J. M. and Griffiths, P. R. (Eds), Wiley, Chichester, UK, 2002, pp. 2993–3004.
9. Li-Chen, E. C. Y., Ismail, A. A., Sedman, J. and van de Voort, F. R., 'Vibrational Spectroscopy of Food and Food Products', in *Handbook of Vibrational Spectroscopy*, Vol. 5, Chalmers, J. M. and Griffiths, P. R. (Eds), Wiley, Chichester, UK, 2002, pp. 3629–3662.
10. Anderson, S. K., Hansen, P. W. and Anderson, H. V., 'Vibrational Spectroscopy in the Analysis of Dairy Products and Wine', in *Handbook of Vibrational Spectroscopy*, Vol. 5, Chalmers, J. M. and Griffiths, P. R. (Eds), Wiley, Chichester, UK, 2002, pp. 3672–2682.
11. Meurens, M. and Yan, S. H., 'Applications of Vibrational Spectroscopy in Brewing,' in *Handbook of Vibrational Spectroscopy*, Vol. 5, Chalmers, J. M. and Griffiths, P. R. (Eds), Wiley, Chichester, UK, 2002, pp. 3663–3671.
12. McClure, W. F. and Stanfield, D. L., 'Near-infrared Spectroscopy of Biomaterials', in *Handbook of Vibrational Spectroscopy*, Vol. 1, Chalmers, J. M. and Griffiths, P. R. (Eds), Wiley, Chichester, UK, 2002, pp. 212–225.
13. Osborne, B. G. and Fearn, T., *Near-Infrared Spectroscopy in Food Analysis*, Wiley, New York, 1988.
14. O'Sullivan, A., O'Connor, B., Kelly, A. and McGrath, M. J., *Int. J. Dairy Technol.*, **52**, 139–148 (1999).
15. Williams, P., 'Near-infrared Spectroscopy of Cereals', in *Handbook of Vibrational Spectroscopy*, Vol. 5, Chalmers, J. M. and Griffiths, P. R. (Eds), Wiley, Chichester, UK, 2002, pp. 3693–3719.
16. Leclerc, D. F. and Trung, T. P., 'Vibrational Spectroscopy in the Pulp and Paper Industry', in *Handbook of Vibrational Spectroscopy*, Vol. 4, Chalmers, J. M. and Griffiths, P. R. (Eds), Wiley, Chichester, UK, 2002, pp. 2952–2976.
17. Leclerc, D. F., 'Fourier-Transform Infrared Spectroscopy in the Pulp and Paper Industry', in *Encyclopedia of Analytical Chemistry*, Vol. 10, Meyers, R. A. (Ed.), Wiley, Chichester, UK, 2000, pp. 8361–8388.
18. Carr, C., 'Vibrational Spectroscopy in the Paint Industry', in *Handbook of Vibrational Spectroscopy*, Vol. 4, Chalmers, J. M. and Griffiths, P. R. (Eds), Wiley, Chichester, UK, 2002, pp. 2935–2951.
19. Adamsons, K., *Prog. Polym. Sci.*, **25**, 1363–1409 (2000).
20. Urban, M. W., 'Infrared and Raman Spectroscopy and Imaging in Coatings Analysis', in *Encyclopedia of Analytical Chemistry*, Vol. 3, Meyers, R. A. (Ed.), Wiley, Chichester, UK, 2000, pp. 1756–1773.

21. Ryland, S. G., 'Infrared Microspectroscopy of Forensic Paint Evidence', in *Practical Guide to Infrared Microspectroscopy*, Humecki, H. J. (Ed.), Marcel Dekker, New York, 1995, pp. 163–243.
22. *An Infrared Spectroscopy Atlas for the Coatings Industry*, 4th Edn, Federation of Societies for Coatings Technology, Philadelphia, PA, USA, 1991.
23. Hummel, D. O., *Atlas of Polymer and Plastics Analysis*, Carl Hanser Verlag, Munich, Germany, 1991.
24. Derrick, M. R., Stulik, D. and Landry, J. M., *Infrared Spectroscopy in Conservation Science*, Getty Conservation Institute, Los Angeles, CA, USA, 1999.
25. Derrick, M. R., 'Infrared Microspectroscopy in the Analysis of Cultural Artifacts', in *Practical Guide to Infrared Microspectroscopy*, Humecki, H. J. (Ed.), Marcel Dekker, New York, 1995, pp. 287–322.
26. Newman, R., *J. Am. Institute Conservation*, **19**, 42–62 (1980).
27. Feller, R. L. (Ed.), *Artists' Pigments: A Handbook of Their History and Characteristics*, Vol. 1, Cambridge University Press, Cambridge, UK, 1986.
28. Roy, A. (Ed.), *Artists' Pigments: A Handbook of Their History and Characteristics*, Vol. 2, National Gallery of Art, Washington, DC, USA, 1993.
29. West-Fitzhugh, E. (Ed.), *Artists' Pigments: A Handbook of Their History and Characteristics*, Vol. 3, National Gallery of Art, Washington, DC, USA, 1997.
30. Visser, T., 'Infrared Spectroscopy in Environmental Analysis', in *Encyclopedia of Analytical Chemistry*, Vol. 1, Meyers, R. A. (Ed.), Wiley, Chichester, UK, 2000, pp. 1–21.
31. Griffith, D. W. T., 'FTIR Measurements of Atmospheric Trace Gases and their Fluxes', in *Handbook of Vibrational Spectroscopy*, Vol. 4, Chalmers, J. M. and Griffiths, P. R. (Eds), Wiley, Chichester, UK, 2002, pp. 2823–2841.
32. Griffith, D. W. T. and Jamie, I. M., 'Fourier-Transform Infrared Spectrometry in Atmospheric and Trace Gas Analysis', in *Encyclopedia of Analytical Chemistry*, Vol. 3, Meyers, R. A. (Ed.), Wiley, Chichester, UK, 2000, pp. 1979–2007.
33. Yokelson, R. J. and Bertschi, I. T., 'Vibrational Spectroscopy in the Study of Fires', in *Handbook of Vibrational Spectroscopy*, Vol. 4, Chalmers, J. M. and Griffiths, P. R. (Eds), Wiley, Chichester, UK, 2002, pp. 2879–2886.
34. Kawaguchi, K., 'Astronomical Vibrational Spectroscopy', in *Handbook of Vibrational Spectroscopy*, Vol. 4, Chalmers, J. M. and Griffiths, P. R. (Eds), Wiley, Chichester, UK, 2002, pp. 2803–2822.
35. Garizi, N., Macias, A., Furch, T., Fan, R., Wagenknecht, P. and Singmaster, K. A., *J. Chem. Edu.*, **78**, 1665–1666 (2001).
36. Kalasinsky, K. S., *J. Chromatogr. Sci., A*, **21**, 246–253 (1983).

Responses to Self-Assessment Questions

Chapter 1

Response 1.1

(i) $\lambda = 1/\bar{\nu} = 1/1656 \ \text{cm}^{-1} = 6.039 \times 10^{-4} \ \text{cm} = 6.039 \ \mu\text{m}$

(ii) $\nu = c/\lambda = 2.998 \times 10^8 \ \text{m s}^{-1}/6.039 \times 10^{-6} \ \text{m} = 4.964 \times 10^{13} \ \text{Hz}$

(iii) $\Delta E = h\nu = 6.626 \times 10^{-34} \ \text{J s} \times 4.964 \times 10^{13} \ \text{Hz} = 3.107 \times 10^{-20} \ \text{J}$

Response 1.2

Chloroform has five atoms and is non-linear and, hence, has $3 \times 5 - 6 = 9$ vibrational degrees of freedom.

Response 1.3

Using Equation (1.9):

$$\bar{\nu} = (1/2\pi c)\sqrt{(k/\mu)}$$

and assuming that k is the same for both bonds, then the ratio of the reduced masses needs only be calculated. For C–H, we have:

$\mu = m_1 m_2/(m_1 + m_2)$

$\quad = (1.993 \times 10^{-27} \ \text{kg} \times 1.674 \times 10^{-27} \ \text{kg})/(1.993 \times 10^{-27} \ \text{kg} + 1.674 \times 10^{-27} \ \text{kg})$

$\quad = 1.545 \times 10^{-27} \ \text{kg}$

Infrared Spectroscopy: Fundamentals and Applications   B. Stuart
© 2004 John Wiley & Sons, Ltd   ISBNs: 0-470-85427-8 (HB); 0-470-85428-6 (PB)

For C–D:

$$\mu = (1.993 \times 10^{-27} \text{ kg} \times 3.345 \times 10^{-27} \text{ kg})/(1.993 \times 10^{-27} \text{ kg} + 3.345 \times 10^{-27} \text{ kg})$$

$$= 2.864 \times 10^{-27} \text{ kg}$$

The vibrational wavenumber is inversely proportional to the square root of these values:

$$\bar{v}_D/\bar{v}_H = \sqrt{\mu_H}/\sqrt{\mu_D} = (\sqrt{1.545 \times 10^{-27} \text{ kg}})/(\sqrt{2.864 \times 10^{-27} \text{ kg}}) = 0.73$$

Thus, if the C–H stretching for $CHCl_3$ is at 3000 cm^{-1}, the C–D stretching is expected at:

$$3000 \times 0.73 = 2190 \text{ cm}^{-1}$$

Response 1.4

The illustrated CO_2 molecule has no dipole moment, but in the bent molecule shown in Figure 1.12 there is a net dipole in the direction shown. Hence, the vibration is 'infrared-active'.

Response 1.5

The first overtone will occur at double the wavenumber of the fundamentals, and so bands would be expected at 1460, 1800 and 5900 cm^{-1}. The possible combinations are at:

$$730 \text{ cm}^{-1} + 1400 \text{ cm}^{-1} = 2130 \text{ cm}^{-1}$$

$$730 \text{ cm}^{-1} + 2950 \text{ cm}^{-1} = 3680 \text{ cm}^{-1}$$

$$1400 \text{ cm}^{-1} + 2950 \text{ cm}^{-1} = 4350 \text{ cm}^{-1}$$

Chapter 2

Response 2.1

(a) $\delta = 100$ mm $= 10$ cm
and therefore the limiting resolution $= 1/10$ cm $= 0.1$ cm^{-1}

(b) $\delta = 1/0.02$ cm^{-1} $= 50$ cm $= 500$ mm

Response 2.2

Most of the materials listed in Table 2.1 are soluble in water, and so the two options available for an aqueous solution are CaF_2 or BaF_2. However, neither of these materials can be used for solutions at extreme pH values, and so the pH should be maintained at around 7.

Response 2.3

The crystal size of the sample is too large, which leads to a scattering of radiation that becomes worse at the high-wavenumber end of the spectrum. Additionally, the bands are distorted and consequently their positions are shifted, thus leading to the possibility of an incorrect assignment. In this case, the sample needs to be ground further with a mortar and pestle.

Response 2.4

The number of peak-to-peak fringes must be counted. For instance, the $3780-1180$ cm^{-1} region may be chosen, since both of these wavenumbers correspond to the tops of peaks. In this range, there are 13 fringes, and so the pathlength of the cell, L, is calculated as follows:

$$L = \frac{n}{2(\bar{v}_1 - \bar{v}_2)}$$

$$= 13/2 \times 2600 \text{ cm}^{-1} = 2.5 \times 10^{-3} \text{ cm}$$

Response 2.5

(a) First, convert the wavenumber values into wavelength by using the relationship $\lambda = 1/\bar{v}$:

$$\lambda = 1/1000 \text{ cm}^{-1} = 10^{-3} \text{ cm} = 10^{-5} \text{ m}$$

$$d_p = (10^{-5} \text{ m}/1.5)/2\pi[\sin(60°) - (1.5/2.4)^2]^{1/2}$$

$$= 1.5 \times 10^{-6} \text{ m}$$

$$= 1.5 \text{ μm}$$

(b) $d_p = 1.0$ μm

(c) $d_p = 0.5$ μm

The depth of penetration at higher wavenumbers (3000 cm^{-1}) is notably less than at lower wavenumbers (1000 cm^{-1}).

Response 2.6

(a) DRIFT is a particularly useful technique for powders, and the sample may be examined by using KBr.

(b) Such a small quantity of sample requires a microsampling accessory. If a microscope attachment to the FTIR spectrometry is not available, a diamond anvil cell (DAC) may be used for samples of microgram quantities.

(c) In order to identify a mixture of amphetamines, GC–IR spectroscopy should be employed. This method allows the components to be separated and the spectra to be differentiated.

Chapter 3

Response 3.1

The intensity of an infrared absorption depends on the change in dipole moment during the vibration. When a hydrogen atom is held in a hydrogen bond, it is in a much more polar environment than when unchelated. The hydrogen atom is moving between two polar centres. It is therefore reasonable to expect an increase in intensity from dimeric and higher polymeric species.

Response 3.2

1.0% (wt/vol) means 1.0 g is dissolved in 100 cm^3 (10 gl^{-1}). The molecular weight of hexanol ($C_6H_{13}OH$) is 92 gmol^{-1}, and so the concentration is 10 gl^{-1}/92 gmol^{-1} = 0.11 moll^{-1}. From the Beer–Lambert law (Equation (3.1)):

$$\varepsilon = A/cl = 0.37/(0.11 \text{ mol} l^{-1} \times 1.0 \text{ mm})$$

$$\varepsilon = 3.4 \text{ l} mol^{-1} mm^{-1}$$

Response 3.3

First, a calibration graph should be produced by using the information provided by the infrared spectra of the standard solutions. A plot of absorbance versus concentration for the caffeine standards is shown in Figure SAQ 3.3. An absorbance of 0.166 corresponds to a concentration of 9 mg ml^{-1}. Thus, the unknown solution contains 9 mg ml^{-1} of caffeine.

Response 3.4

The first step is to draw a calibration plot of absorbance against concentration and this is shown in Figure SAQ 3.4. The slope of this plot is 5.10/(vol%). 1 vol% means that there is 1 cm^3 in 100 cm^3, and so 10 cm^3 in 1 l. 10 cm^3 has a mass:

$$m = \rho V = 0.790 \text{ g} cm^{-3} \times 10 \text{ cm}^3 = 7.90 \text{ g}$$

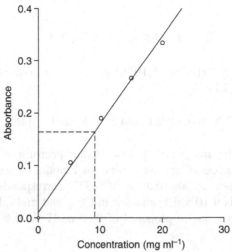

Figure SAQ 3.3 Calibration graph for caffeine solutions. From Stuart, B., *Biological Applications of Infrared Spectroscopy*, ACOL Series, Wiley, Chichester, UK, 1997. © University of Greenwich, and reproduced by permission of the University of Greenwich.

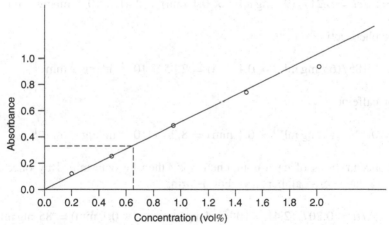

Figure SAQ 3.4 Calibration graph for the determination of acetone. From Stuart, B., *Modern Infrared Spectroscopy*, ACOL Series, Wiley, Chichester, UK, 1996. © University of Greenwich, and reproduced by permission of the University of Greenwich.

The molecular weight of acetone is 58 $g\,mol^{-1}$, and so the equivalent concentration is:

$$7.90\ g\,l^{-1}/58\ g\,mol^{-1} = 0.136\ mol\,l^{-1}$$

Thus:

$$1 \text{ vol}\% = 0.136 \text{ mol } l^{-1}$$

and so the slope $= 5.10/(\text{vol}\%)$ $5.10/0.136 \text{ mol } l^{-1} = 37.5 \text{ l mol}^{-1}$
From Equation (3.1):

$$\varepsilon = \text{slope}/l = 37.5 \text{ l mol}^{-1}/0.1 \text{ mm} = 375 \text{ l mol}^{-1} \text{ mm}^{-1}$$

The next stage in the analysis is to look at the spectrum of the unknown commercial sample of propanol and determine the amount of acetone present. From the calibration graph, an absorbance of 0.337 corresponds to 0.67 vol%, or 0.09 mol l^{-1}. This is a 10 vol% solution in CCl_4 and multiplication by 10 gives a concentration of acetone in propanol of 0.90 mol l^{-1}, or 6.7 vol%.

Response 3.5

The Beer–Lambert law (Equation (3.1)) is used to determine the molar absorptivity for each component, $A = \varepsilon cl$. For aspirin:

$$\varepsilon = A/cl = 0.217/(90 \text{ mg ml}^{-1} \times 0.1 \text{ mm}) = 2.41 \times 10^{-2} \text{ ml mg}^{-1} \text{ mm}^{-1}$$

For phenacetin:

$$\varepsilon = 0.185/(65 \text{ mg ml}^{-1} \times 0.1 \text{ mm}) = 2.85 \times 10^{-2} \text{ ml mg}^{-1} \text{ mm}^{-1}$$

For caffeine:

$$\varepsilon = 0.123/(15 \text{ mg ml}^{-1} \times 0.1 \text{ mm}) = 8.20 \times 10^{-2} \text{ ml mg}^{-1} \text{ mm}^{-1}$$

The concentrations of each component may then be determined by once again applying the Beer–Lambert law. For aspirin:

$$c = A/\varepsilon l = 0.207/(2.41 \times 10^{-2} \text{ ml mg}^{-1} \text{ mm}^{-1} \times 0.1 \text{ mm}) = 85 \text{ mg ml}^{-1}$$

For phenacetin:

$$c = 0.202/(2.85 \times 10^{-2} \text{ ml mg}^{-1} \text{ mm}^{-1} \times 0.1 \text{ mm}) = 71 \text{ mg ml}^{-1}$$

For caffeine:

$$c = 0.180/(8.20 \times 10^{-2} \text{ ml mg}^{-1} \text{ mm}^{-1} \times 0.1 \text{ mm}) = 22 \text{ mg ml}^{-1}$$

Chapter 4

Response 4.1

The band marked A in Figure 4.1 shows four overlapped bands at around 2960, 2930, 2870 and 2850 cm^{-1}. Those at 2960 and 2870 cm^{-1} are the asymmetric and symmetric stretching bands of the methyl group, respectively. The others are the corresponding bands for the methylene group. The band marked B shows two overlapped components at 1465 and 1450 cm^{-1}. The higher-wavenumber band is the CH_2 bending band, while the lower is the asymmetric CH_3 bending band. The band marked C absorbs at 1380 cm^{-1} and is a symmetrical CH_3 bending band.

Response 4.2

2-Methylbutane has the structure $(CH_3)_2CHCH_2CH_3$ and contains a dimethyl group. It does not contain a chain of four CH_2 groups and is therefore in category (b). Cyclopentane is a five-membered ring, containing a chain of five CH_2 groups, and so it will absorb at 720 cm^{-1} and is in category (a). 2-Methyloctane has the structure $(CH_3)_2CH(CH_2)_5CH_3$ and contains both a dimethyl group and a chain of five CH_2 groups, and so is in category (c). 3-Methylpentane has the structure $CH_3CH_2CH(CH_3)CH_2CH_3$ and does not contain either feature, and so is in category (d). Butane has the structure $CH_3CH_2CH_2CH_3$ and is in category (b). 1,1-Dimethylcyclohexane contains both structural features and absorbs at both 720 and 1375 cm^{-1}, and is in category (c).

Response 4.3

These compounds are very easily differentiated by using the C–H deformation bands between 600 and 1000 cm^{-1}. Compound A shows one band at 734 cm^{-1}, compound B shows one band at 787 cm^{-1}, while compound C shows two bands at 680 and 760 cm^{-1}. Use of Figure 4.2 suggests that the only disubstitution pattern giving two peaks in this region is the 1,3-substituted compound. It also suggests that the 1,2- and 1,4-species give one peak, with the 1,2-compound absorbing at a lower wavenumber than the 1,4-material. Hence, A is 1,2-disubstituted, B is 1,4-disubstituted and C is 1,3-disubstituted.

Response 4.4

The 1815 and 1750 cm^{-1} pair corresponds to the C=O stretching bands of the compound shown in Figure 4.7(a), while the 1865 and 1780 cm^{-1} pair is associated with the compound shown in Figure 4.7(b). The latter compound is cyclic, and so the bands produced by this compound will appear at higher wavenumbers.

Response 4.5

In Figure 4.8, there is a doublet with peaks at 3380 and 3300 cm^{-1}, characteristic of a primary amine. There is also a series of overlapped bands below 3000 cm^{-1}

due to C–H stretching. The band near 1600 cm^{-1} may be assigned to NH$_2$ scissoring and N–H bending. The band near 1450 cm^{-1} is due to C–H bending. Aliphatic C–N stretching appears at 1220–1020 cm^{-1} and a broad band at 850–750 cm^{-1} is due to NH$_2$ wagging and twisting. N–H wagging appears at 715 cm^{-1}.

Response 4.6

The broad bands appearing at wavenumbers above 1500 cm^{-1} are associated with hydrogen bonding and may be attributed to a P(O)OH group. The sharper band at 1300 cm^{-1} can be assigned to P–CH$_3$ stretching. The broad band at 1200 cm^{-1} indicates a P=O unit and confirms the presence of the P(O)OH group. In addition, the presence of a band at 1055 cm^{-1} suggests that the molecule contains a P–O–CH$_3$ group. The absence of sharp bands in the 2500–2000 cm^{-1} range shows that a P–H group cannot be present. Combining this information gives us a phosphoric acid with the structure shown in Figure SAQ 4.6.

Figure SAQ 4.6 Structure of the phosphoric acid C$_2$H$_7$PO$_3$.

Response 4.7

There is a series of bands in the 2150–2180 nm region and a strong sharp peak at 2460 nm. These bands are combination bands. The 2150–2180 nm bands are combinations of C–H and C–C stretching, while the 2460 nm band is a combination of C–H stretching and C–H bending. The spectrum also shows bands near 1670 nm due to the first overtone of C–H stretching. The second overtone at 1130 nm is very weak in this spectrum.

Response 4.8

Inspection of the high-wavenumber end of the infrared spectrum of vanillin shows a weak band at 3550 cm^{-1} which could be due to an alcohol or phenol group. In the 3100–2700 cm^{-1} range, there are a number of C–H stretching bands. First, there are bands above and below 3000 cm^{-1}, which indicate the presence of both aromatic and aliphatic C–H groups. The C–H stretching bands at 2800 and 2700 cm^{-1} are lower in wavenumber, so indicating the presence of an aldehyde group. The carbonyl stretching bands near 1790 and 1740 cm^{-1} can be assigned to non-hydrogen-bonded and hydrogen-bonded aldehyde C=O stretching, respectively. Thus, it has been established that vanillin contains a benzene ring, an O–H group, both aromatic and aliphatic C–H groups, and an aldehyde group. The actual structure of vanillin is shown in Figure SAQ 4.8.

CHO

OCH$_3$

OH

Figure SAQ 4.8 Structure of vanillin.

Response 4.9

Looking first at wavenumbers greater than 1500 cm^{-1}, there are bands at 1520, 1610 and 3480 cm^{-1} and aliphatic and aromatic C–H stretching bands near 3000 cm^{-1}. The sharp band at the highest wavenumber cannot be due to O–H stretching as the sample is a liquid film and the O–H stretching band would be expected to be a broad band due to hydrogen bonding, as well as the fact that the known empirical formula does not contain oxygen. Thus, the band at 3480 cm^{-1} is due to N–H stretching and the compound is an amine. The multiplicity of the band implies that the compound is a *secondary* amine. The band at 1610 cm^{-1} could be due to N–H bending, but is more likely to be due to benzene ring vibrations, or both. The band at 1520 cm^{-1} is due to an aliphatic bending band, confirmed by a doublet below 3000 cm^{-1}. The spectrum below 1500 cm^{-1} supports the idea of a monosubstituted benzene ring. Thus, the spectrum supports the fact that the compound is *N*-methylaniline (C$_6$H$_5$–NH–CH$_3$).

Response 4.10

A notable characteristic of the spectrum shown in Figure 4.15 is its simplicity. The C–H stretching bands confirm that no unsaturation is present, since there are no bands above 3000 cm^{-1}. The band at 1467 cm^{-1} is the scissoring frequency of CH$_2$ groups, while that at 1378 cm^{-1} is the symmetrical bending mode of a CH$_3$ group. The absence of bands between 1300 and 750 cm^{-1} suggests a straight-chain structure, while the band at 782 cm^{-1} tells us that there are four or more CH$_2$ groups in the chain. The compound is, in fact, *n*-decane.

Response 4.11

This is a saturated molecule, as there are no C–H stretching bands above 3000 cm^{-1}. The broad band between 3700 and 3200 cm^{-1} tells us that this compound is an alcohol or a phenol. An alcohol containing four carbon atoms fits the molecular formula. There is no band at 720 cm^{-1}, and so the compound must be branched. There is also a doublet at 1386 and 1375 cm^{-1}, hence indicating an isopropyl group. The very strong band at 1040 cm^{-1} is due to C–O stretching. The compound must therefore be 2-methylpropan-1-ol, (CH$_3$)$_2$CHCH$_2$OH.

Response 4.12

The formula C_8H_{16} suggests that there must be one C=C double bond in the molecule, since a fully saturated molecule would have the formula C_8H_{18}. The band above 3000 cm^{-1} in Figure 4.17 suggests that there are hydrogens attached to a double bond. The spectrum also shows a C=C stretching wavenumber at 1650 cm^{-1} and C–H deformation bands at 998 and 915 cm^{-1}, with a weak band at 720 cm^{-1}. The C=C stretching band is intense and is, therefore, not a *trans*-alkene, with the double bond far from the end of the chain. The band at 720 cm^{-1} gives similar information and suggests at least four CH$_2$ groups. The C–H deformation bands suggest that the molecule contains the group –CH=CH$_2$. The only possible structure is, therefore, oct-1-ene.

Response 4.13

The molecule is aromatic, and there are C–H stretching bands above 3000 cm^{-1}. The small bands just below 3000 cm^{-1} are probably overtones from the very strong bands at 1530 and 1350 cm^{-1}. The bands at 850 and 700 cm^{-1} suggest that this is a mono-substituted benzene. The two very strong bands at 1530 and 1350 cm^{-1} tell us that a NO$_2$ group is present. The liquid is, therefore, nitrobenzene.

Response 4.14

The strong doublet at 3380 and 3180 cm^{-1} suggests an NH$_2$ group. There is also a C=O stretching band at 1665 cm^{-1}. The molecule appears to be a primary amide, containing a benzene ring. The molecule is actually benzamide.

Chapter 5

Response 5.1

Figure 5.1 shows a broad band centred near 1450 cm^{-1} and a sharp band near 800 cm^{-1}. Consultation with Table 5.1 reveals that the most likely anion in $CO_3{}^{2-}$. The spectrum is, in fact, that of calcium carbonate.

Response 5.2

Anhydrous CaSO$_4$ shows bands due to S–O stretching at 1140–1080 cm^{-1} and S–O bending at 620 cm^{-1}. The hemihydrate CaSO$_4$ spectrum shows these S–O bands, but also asymmetric and symmetric O–H stretching bands in the 3700–3200 cm^{-1} region and a single peak near 1600 cm^{-1} due to O–H bending. These O–H bands indicate that the water molecules are subjected to one type of environment in the lattice. The spectrum of dihydrate $CaSO_4$ shows the same S–O bonds, but shows a more complex O–H stretching region and a doublet in the 1700–1600 cm^{-1} region. The O–H stretching region consists of four overlapped components, representing asymmetric and symmetric stretching in two different environments. The splitting of the O–H bending band also reflects different environments for the water molecules in the crystal lattice structure. Thus, the degree of hydration may be determined for $CaSO_4$ based on the appearance of water bands in the infrared spectrum.

Response 5.3

All of the major bands of ethylenediamine listed in Table 5.7 are shifted to lower-wavenumber values when the ligand is complexed with zinc. The shifts may be attributed to the weakening of the N–H bonds, resulting from the drainage of electrons from the nitrogen atom when the metal ion is coordinated.

Response 5.4

The $Ru_3(CO)_{12}$ compound shows three bands in the 2000 cm^{-1} region and none below 1900 cm^{-1}, so indicating the presence of bridging ligands and no terminal carbonyls. By comparison, $Fe_3(CO)_{12}$ shows carbonyl bands above and below 1900 cm^{-1}, thus indicating that this iron complex has bridging CO ligands as well as terminal ligands.

Response 5.5

Figure 5.9 shows four infrared bands in the O–H stretching region. The 3620 cm^{-1} band is due to the inner hydroxyl groups lying between the tetrahedral and octahedral sheets. The bands at 3695, 3669 and 3653 cm^{-1} are due to stretching vibrations of the O–H groups at the octahedral surface of the layers. There are also bands due to Si–O stretching bands in the 1120–1000 cm^{-1} range. The spectrum also shows bands at 936 and 914 cm^{-1} which are due to O–H bending bands. Each of these observations indicates that kaolinite is a clay that has most of its octahedral sites occupied by trivalent ions. As the O–H bending bands appear in the 950–800 cm^{-1} region, this spectrum indicates that kaolinite has trivalent central atoms in the octahedral sheets. The bands at 1120–1000 cm^{-1} and the four bands in the 3800–3400 cm^{-1} region are characteristic of a clay with trivalent atoms in the octahedral sheets.

Chapter 6

Response 6.1

The major infrared modes of poly(*cis*-isoprene) observed in Figure 6.4 are listed in Table SAQ 6.1. These assignments are made as a result of consultation with the correlation table shown in Figure 6.1.

Table SAQ 6.1 Major infrared bands of poly(*cis*-isoprene)

Wavenumber (cm^{-1})	Assignment
3036	=C–H stretching
2996–2855	CH$_2$; CH$_3$ stretching
1663	C=C stretching
1452	C–H bending
1376	C–H bending
837	C–H out-of-plane bending

Response 6.2

In Figure 6.6, there are bands in the C–H stretching region, above and below 3000 cm^{-1}, and so the molecule must contain both aromatic and aliphatic C–H groups. Typical benzene ring absorptions are observed, i.e. C–H stretching at 3100–3000 cm^{-1}, overtone and combination bands at 2000–1650 cm^{-1}, ring stretching at 1600–1550 cm^{-1}, ring stretching at 1500–1450 cm^{-1}, C–H in-plane bending at 1300–1000 cm^{-1}, and C–H out-of-plane bending at 900–600 cm^{-1}. The two peaks at 700 and 780 cm^{-1} indicate a mono-substituted ring. A benzene ring is clearly the main functional group in this polymer, which is polystyrene.

Response 6.3

One of the most notable features of the spectrum shown in Figure 6.7 is a very intense broad band in the 1100–1000 cm^{-1} range, which can be readily assigned to the asymmetric Si–O–Si stretching of a siloxane. The intense band between 1300 and 1200 cm^{-1} indicates the presence of a Si–CH$_3$ group – this band is due to symmetric CH$_3$ bending. There is also evidence of silicon attached to a benzene ring: bands near 1430 and 1110 cm^{-1} are due to Si–C$_6$H$_5$ stretching. The backbone of a silicon-based polymer must consist of a Si–O–Si backbone and it is known that phenyl and methyl groups are attached to the silicon. The spectrum is of a polymethylphenylsiloxane.

Response 6.4

A plot of the absorbance ratio 1020 cm^{-1}/720 cm^{-1} as a function of concentration is linear and can be used as a calibration plot, and this is presented in Figure SAQ 6.4. The absorbance ratio for the unknown sample is 0.301/0.197 = 1.53 and consultation with the calibration graph gives a concentration of 17% ethyl acetate for the copolymer. Thus, the composition of the copolymer is 83% ethylene/17% ethyl acetate.

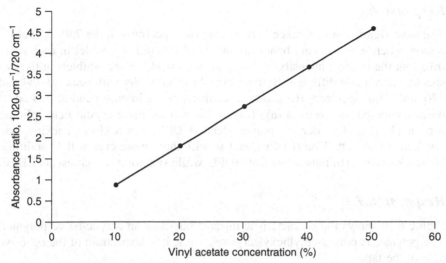

Figure SAQ 6.4 Calibration plot for ethylene/vinyl acetate copolymers.

Response 6.5

The data presented in Table 6.4 can be used to produce a calibration curve. Figure SAQ 6.5 shows a plot of the 2270 cm^{-1} band absorbance versus the isocyanate content for the known samples. The absorbance at 2270 cm^{-1} for the unknown sample is 0.254 – corresponding to a content of 7.6%.

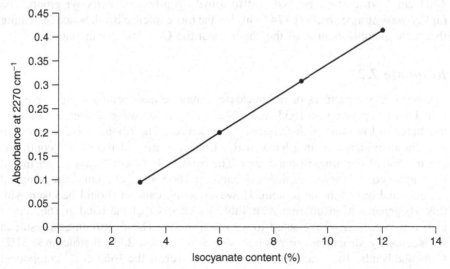

Figure SAQ 6.5 Calibration plot for isocyanate in polyurethane prepolymers.

Response 6.6

The most significant difference between the two spectra is in the $750-700$ cm^{-1} region, where CH$_2$ rocking bands are observed. There is a doublet in each spectrum, but the relative intensity of the component peaks of the doublets in the two spectra are notably different. Both spectra show a doublet with peaks at 734 and 720 cm^{-1}, but spectrum (b) shows a relatively more intense band at 734 cm^{-1} when compared to spectrum (a). This is evidence of more crystal field splitting for sample (b). High-density polyethylene (HDPE) has a closer packing than low-density polyethylene (LDPE), and so will show more crystal field splitting. Hence, spectrum (b) must represent HDPE, while spectrum (a) represents LDPE.

Response 6.7

Figure 6.18 shows the characteristic infrared bands of an acrylic-based polymer. The polymer is poly(2-ethylhexylacrylate). Thus, it is a spectrum of the adhesive side of the tape.

Chapter 7

Response 7.1

The band at 1624 cm^{-1} in Figure 7.3 is due to the asymmetrical CO$_2^-$ stretching vibration of phosphatidylserine. The band at 1737 cm^{-1} is broad, consisting of two ester C=O stretching bands at 1742 and 1728 cm^{-1} and a small band at 1701 cm^{-1} which may be assigned to a hydrogen-bonded carbonyl group. The higher-wavenumber band at 1742 cm^{-1} is the more intense band, hence indicating that a *trans*-conformation of the chain about the C–C bond is favoured.

Response 7.2

The secondary structures of ribonuclease S may be assigned by using Table 7.4. The bands appearing at 1633 and 1672 cm^{-1} are at wavenumbers commonly attributed to low- and high-frequency β-structures. The ribonuclease S spectrum also shows evidence of an α-helix, with a band appearing at 1653 cm^{-1} contributing to 25% of the amide I band area. The band at 1645 cm^{-1} may be attributed to random coils. The two remaining bands at 1663 and 1681 cm^{-1} are assigned to turns and bends in the protein. However, some caution should be taken with this assignment. Consultation with Table 7.4 shows that the band at 1663 cm^{-1} also lies in the range attributed to a '3-turn' helix. These assignments result in the secondary structures for ribonuclease S: α-helices, 25%; β-structures, 31%; turns and bends, 16%; random coils, 8%. However, if the 1663 cm^{-1} component is assigned to a '3-turn' helix, this would result in 34% helical structure, which is higher than expected. X-ray crystallography and circular dichroism studies of the structure of ribonuclease S show that the protein contains about 25% helix. Thus, this example emphasizes the need to exercise caution in making absolute quantitative determinations.

Response 7.3

The component band at 1692 cm^{-1} may be attributed to turns and bends in the peptide. Computer analysis shows that this band contributes to 11% to the total amide I band area. The band at 1675 cm^{-1} (7% of the total band area) appears at a wavenumber commonly expected for a β-structure. The large component at 1653 cm^{-1} appears at a wavenumber that may be attributed to unordered structures in the peptide. This band is the major component, forming 82% of the band area. Most small peptides show largely random structures, but this is not surprising because of the small size of such molecules, which limits the possibilities of forming structures such as α-helices which require a minimum number of amino acids to form.

Response 7.4

Figure 7.12 shows that the non-cancerous tissue gives rise to a maximum at 1162 cm^{-1} and a smaller shoulder near 1172 cm^{-1} in the C–O stretching band. The tissue containing 80% malignancy, however, shows a maximum at 1172 cm^{-1} and a weak shoulder at the lower wavenumber. The C–O stretching band intensities of the 20% cancerous tissue is intermediate between these values. The observed changes in the spectra arc due to changes in the degree of hydrogen bonding associated with the C–O groups in the tissue (belonging to cell proteins). Hydrogen bonding affects the wavenumbers of the modes involved in such bonds. The lower-wavenumber component observed in these spectra can be attributed to hydrogen-bonded C–O groups, whereas the higher-wavenumber component is due to non-hydrogen-bonded C–O groups. Thus, the significant decrease of the intensity of the low-wavenumber component band in malignant tissues suggest that in colon cancer most of the hydrogen bonds in the C–O groups have disappeared.

Chapter 8

Response 8.1

At the higher-wavenumber end of the spectrum shown in Figure 8.5, there are two O–H stretching bands at 3525 and 3370 cm^{-1}, representing different types of hydrogen bonding for the molecule. A variation in the hydrogen-bonding environment is also detected in the C=O stretching region: there are three distinct bands at 1669, 1660 and 1649 cm^{-1}. A C=C stretching band appears at

1615 cm^{-1}, while C–H stretching bands appear in the $3000–2800$ cm^{-1} region. The region below 1500 cm^{-1} is quite complex, but some important bands can be assigned: asymmetric C–CH$_3$ bending at 1450 cm^{-1}, symmetric C–CH$_3$ bending at 1367 cm^{-1}, O–H bending at 1238 cm^{-1}, C–O stretching at 1070 and 1060 cm^{-1}, and C=CH twisting at 873 cm^{-1}. It should be noted that the infrared spectra of steroids can be difficult to unambiguously assign because minimal functional groups are present and these are often involved in coupled vibrations.

Response 8.2

Polymorph I shows bands at 1760 and 1645 cm^{-1} in the carbonyl region (Figure 8.9(a)). The spectrum of polymorph II (Figure 8.9(b)) shows bands at 1740, 1710 and 1670 cm^{-1}. The carbonyl band is sensitive to hydrogen bonding and the differences in the spectra of the two polymorphs may be attributed to differences in the type of hydrogen bonding in the crystalline structures. In order to carry out quantitative measurements of polymorph II in a sample containing a mixture of polymorphs, measurements of the band areas of the 1670 cm^{-1} carbonyl component band for a series of calibration samples could be carried out. The 1670 cm^{-1} band is suitable for this because it does not overlap with other carbonyl component bands.

Response 8.3

Although ephedrine and pseudoephedrine are stereoisomers, there are differences in the $1300–1000$ cm^{-1} regions of the spectra. These overlapped regions make it difficult to make simple assignments, although they do provide useful fingerprint patterns which are characteristic for each of these stereoisomers.

Response 8.4

Consultation with Figure 8.14 informs us that the intensity of the second derivative spectra increases with alcohol content. The same trend would be expected for the spectra shown in Figure 8.15. Thus, the second derivative spectrum of weakest intensity corresponds to the sample containing 17% ethanol (Tequila). The spectra with increasing intensities correspond to the samples of increasing ethanol concentration, with the most intense spectrum corresponding to the 40% sample (Cognac).

Response 8.5

The spectrum illustrated in Figure 8.17 shows a broad band in the $3500–3000$ cm^{-1} region which may be attributed to water O–H stretching. The remaining bands of significant absorbance are due to the acrylic binder in the paint. Acrylics will show infrared bands due to the C=O and C–O units of the ester group. A C=O stretching band appears at 1730 cm^{-1}, and C–O–C bending and C–O bending are observed at 1190 and 1450 cm^{-1}, respectively. There are also a number of bands associated with the C–H group: C–H stretching at 3000–2800 cm^{-1}, CH$_2$ bending at 1485 cm^{-1}, C–CH$_3$ bending at 1300 cm^{-1}, and CH$_2$ bending at 1145 cm^{-1}.

Response 8.6

The spectrum presented in Figure 8.20(a) shows a few small bands, but does show a strong band at 2100 cm^{-1}. This band could be due to C≡N stretching, hence indicating that this spectrum corresponds to the Fe$_4$[Fe(CN)$_6$] structure of the Prussian Blue pigment. The 2100 cm^{-1} band is attributed to the [Fe(C=N)$_6$]$^{3-}$ ion stretching mode. The spectrum given in Figure 8.20(b) shows broad strong bands in the 1150–950 cm^{-1} region which may be attributed to overlapped Si–O–Si and Si–O–Al stretching bands. The sharp peak at 2340 cm^{-1} is a sulfur ion stretching band (this occurs in some natural ultramarines, but is not observed in synthetic ultramarines). Thus, this spectrum corresponds to the Na$_8$(S$_2$)(Al$_6$Si$_6$O$_{24}$) structure of the ultramarine pigment.

Response 8.7

Atmospheric water gives rise to vibration–rotation bands above 3500 cm^{-1}, while methane produces similar bands centred near 3000 cm^{-1}. The strong peak near 2300 cm^{-1} is due to carbon dioxide, which also shows a band near 600 cm^{-1}. Carbon monoxide shows vibration–rotation bands in the 2200–200 cm^{-1} range. Hydrogen cyanide, a dangerous gas produced in cigarette smoke, is identified by a very sharp peak near 700 cm^{-1}. There is also a band appearing in the spectrum shown in Figure 8.21 which is not attributable to the aforementioned components. A band at 1700 cm^{-1} appears at a wavenumber characteristic of carbonyl stretching and is most likely to be due to acetaldehyde or acetone.

Bibliography

Books

General

Barrow, G. M., *Introduction to Molecular Spectroscopy*, McGraw-Hill, New York, 1962.

Brown, J. M., *Molecular Spectroscopy*, Oxford University Press, Oxford, UK, 1998.

Burns, D. A. and Ciurczak, E. W., *Handbook of Near-Infrared Analysis*, Marcel Dekker, New York, 1992.

Chalmers, J. M. and Griffiths, P. R. (Eds), *Handbook of Vibrational Spectroscopy* (in 5 volumes), Wiley, Chichester, UK, 2002.

Christy, A. A., Ozaki, Y. and Gregoriou, V. G., *Modern Fourier-Transform Infrared Spectroscopy*, Elsevier, Amsterdam, The Netherlands, 2001.

Colthup, N., Daly, L. and Wilberley, S., *Introduction to Infrared and Raman Spectroscopy*, Academic Press, New York, 1990.

Duckett, S. and Gilbert, B., *Foundations of Spectroscopy*, Oxford University Press, Oxford, UK, 2000.

Gauglitz, G., *Handbook of Spectroscopy*, Wiley-VCH, Weinheim, Germany, 2003.

Günzler, H. and Gremlich, H.-U., *IR Spectroscopy: An Introduction*, Wiley-VCH, Weinheim, Germany, 2002.

Hollas, J. M., *Modern Spectroscopy*, Wiley, Chichester, UK, 1996.

Hollas, J. M., *Basic Atomic and Molecular Spectroscopy*, Wiley, Chichester, UK, 2002.

Meyers, R. A. (Ed.), *Encyclopedia of Analytical Chemistry* (in 15 volumes), Wiley, Chichester, UK, 2000.

Infrared Spectroscopy: Fundamentals and Applications   B. Stuart
© 2004 John Wiley & Sons, Ltd   ISBNs: 0-470-85427-8 (HB); 0-470-85428-6 (PB)

Mirabella, F. M., *Modern Techniques in Applied Molecular Spectroscopy*, Wiley, New York, 1998.

Perkampus, H. H., *Encyclopedia of Spectroscopy: A Comprehensive Handbook*, Wiley-VCH, Weinheim, Germany, 1995.

Siesler, H. W., *Near-Infrared Spectroscopy*, Wiley-VCH, Weinheim, Germany, 2002.

Smith, A. L., *Applied Infrared Spectroscopy: Fundamentals, Techniques and Analytical Problem Solving*, Wiley, New York, 1979.

Smith, B. C., *Fundamentals of Fourier-Transform Infrared Spectroscopy*, CRC Press, Boca Raton, FL, USA, 1996.

Theophanides, T. (Ed), *Fourier-Transform Infrared Spectroscopy: Industrial, Chemical and Biochemical Applications*, D. Riedel Publishing Company, Dordrecht, The Netherlands, 1984.

Tranter, G. E. and Holmes, J. L., *Encyclopedia of Spectroscopy and Spectrometry*, Academic Press, New York, 2000.

Workman, J., *Applied Spectroscopy: A Compact Reference for Practitioners*, Academic Press, New York, 1998.

Experimental Methods

Coleman, P. B., *Practical Sampling Techniques for Infrared Analysis*, CRC Press, Boca Raton, FL, USA, 1993.

Davies, S. P., Abrams, M. C. and Brault, J. W., *Fourier-Transform Spectrometry*, Academic Press, New York, 2000.

Humecki, H. J. (Ed.), *Practical Guide to Infrared Microspectroscopy*, Marcel Dekker, New York, 1995.

Kauppinen, J. and Partanen, J., *Fourier Transforms in Spectroscopy*, Wiley-VCH, Weinheim, Germany, 2001.

White, R., *Practical Spectroscopy: Chromatography–Fourier-Transform Infrared Spectroscopy and its Applications*, Marcel Dekker, New York, 1990.

Analysis

George, W. O. and Willis, H. (Eds), *Computer Methods in UV, Visible and IR Spectroscopy*, The Royal Society of Chemistry, Cambridge, UK, 1990.

Griffiths, P. R. and de Haseth, J. A., *Fourier-Transform Infrared Spectrometry*, Wiley, New York, 1986.

Mark, H., *Principles and Practice of Spectroscopic Calibration*, Wiley, New York, 1996.

Messerschmidt, R. G. and Harthcock, M. A. (Eds), *Infrared Microspectroscopy: Theory and Applications*, Marcel Dekker, New York, 1998.

Nyquist, R. A., *Interpreting Infrared, Raman and Nuclear Magnetic Resonance Spectra*, Academic Press, New York, 2001.

Smith, B. C., *Infrared Spectral Interpretation: A Systematic Approach*, CRC Press, Boca Raton, FL, USA, 1999.

Smith, B. C., *Quantitative Spectroscopy: Theory and Practice*, Elsevier, Amsterdam, The Netherlands, 2002.

Socrates, G., *Infrared and Raman Characteristic Group Frequencies*, Wiley, Chichester, UK, 2001.

Wartewig, S., *IR and Raman Spectroscopy: Fundamental Processing*, Wiley-VCH, Weinheim, Germany, 2003.

Organic Molecules

Feinstein, K., *Guide to Spectroscopic Identification of Organic Compounds*, CRC Press, Boca Raton, FL, USA, 1994.

Field, L. D., Sternhell, S. and Kalman, J., *Organic Structures from Spectra*, Wiley, Chichester, UK, 2002.

Linvien, D., *The Handbook of Infrared and Raman Characteristic Frequencies of Organic Molecules*, Academic Press, Boston, MA, USA, 1991.

Mohan, J., *Organic Spectroscopy: Principles and Applications*, CRC Press, Boca Raton, FL, USA, 2002.

Pretsch, E. and Clerc, J. T., *Spectra Interpretation of Organic Compounds*, Wiley-VCH, Weinheim, Germany, 1997.

Roeges, N. P. G., *A Guide to the Complete Interpretation of Infrared Spectra of Organic Structures*, Wiley, Chichester, UK, 1984.

Schrader, B., *Raman/Infrared Atlas of Organic Compounds*, Wiley-VCH, Weinheim, Germany, 1989.

Silverstein, R. M. and Webster, F. X., *Spectrometric Identification of Organic Compounds*, Wiley, New York, 1997.

Workman, J., *Handbook of Organic Compounds: NIR, IR, Raman and UV–VIS Spectra*, Academic Press, New York, 2000.

Inorganic Molecules

Brisdon, A. K., *Inorganic Spectroscopic Methods*, Oxford University Press, Oxford, UK, 1998.

Clark, R. J. H. and Hester, R. E. (Eds), *Spectroscopy of Inorganic Based Materials*, Wiley, New York, 1987.

Ebsworth, E. A. V., Rankin, D. W. H. and Cradock, S., *Structural Methods in Inorganic Chemistry*, Blackwell, Oxford, UK, 1987.

Greenwood, N. N., *Index of Vibrational Spectra of Inorganic and Organometallic Compounds*, Butterworths, London, 1972.

Nakamoto, K., *Infrared and Raman Spectra of Inorganic and Coordination Compounds, Part A, Theory and Applications in Inorganic Chemistry*, Wiley, New York, 1997.

Nakamoto, K., *Infrared and Raman Spectra of Inorganic and Coordination Compounds, Part B, Applications in Coordination, Organometallic and Bioinorganic Chemistry*, Wiley, New York, 1997.

Nyquist, R. A., Putzig, C. L. and Leugers, M. A., *Handbook of Infrared and Raman Spectra of Inorganic Compounds and Organic Salts*, Academic Press, San Diego, CA, USA, 1997.

Ross, S. D., *Inorganic Vibrational Spectroscopy*, Marcel Dekker, New York, 1971.

Polymers

Bower, D. I. and Maddams, W. F., *The Vibrational Spectroscopy of Polymers*, Cambridge University Press, Cambridge, UK, 1989.

Fawcett, A. H., *Polymer Spectroscopy*, Wiley, Chichester, UK, 1995.

Garton, A., *Infrared Spectroscopy of Polymer Blends, Composites and Surfaces*, Carl Hanser Verlag, Munich, Germany, 1992.

Garton, A., *Infrared Spectroscopy of Polymer Blends*, Hanser Gardner Publications, Cincinnati, OH, USA, 1996.

Koenig, J. L., *Spectroscopy of Polymers*, Elsevier, Amsterdam, The Netherlands, 1999.

Schroder, E., Muller, G. and Arndt, K. F., *Polymer Characterization*, Carl Hanser Verlag, Munich, Germany, 1989.

Siesler, H. W. and Holland-Moritz, K., *Infrared and Raman Spectroscopy of Polymers*, Marcel Dekker, New York, 1980.

Zerbi, G., *Modern Polymer Spectroscopy*, Wiley-VCH, Weinheim, Germany, 1999.

Biological Applications

Ciurczak, E. W. and Drennen, J. K., *Pharmaceutical and Medicinal Applications of Near-Infrared Spectroscopy*, Marcel Dekker, New York, 2002.

Clark, R. J. H. and Hester, R. E., *Biomedical Applications of Spectroscopy*, Wiley, Chichester, UK, 1996.

Gremlich, H.-U. and Yan, B., *Infrared and Raman Spectroscopy of Biological Materials*, Marcel Dekker, New York, 2000.

Havel, H. A. (Ed.), *Spectroscopic Methods for Determining Protein Structures in Solution*, Wiley, New York, 1995.

Mantsch, H. H. and Chapman, D. (Eds), *Infrared Spectroscopy of Biomolecules*, Wiley, New York, 1996.

Mantsch, H. H. and Jackson, M. (Eds), *Infrared Spectroscopy: New Tool in Medicine*, International Society for Optical Engineering, Bellingham, WA, USA, 1998.

Raghavaachari, R. (Ed.), *Near-Infrared Applications in Biotechnology*, Marcel Dekker, New York, 2000.

Industrial Applications

Derrick, M. R., Stulik, D. and Landry, J. M., *Infrared Spectroscopy in Conservation Science*, Getty Conservation Institute, Los Angeles, CA, USA, 1999.

Osborne, B. G. and Fearn, T., *Near-Infrared Spectroscopy in Food Analysis*, Wiley, New York, 1988.

Vandeberg, J. T., *An Infrared Spectroscopy Atlas for the Coatings Industry*, Federation of Societies for Coatings Technology, Philadelphia, PA, USA, 1991.

Journals

Analytica Chimica Acta
Analytical Chemistry
Applied Spectroscopy
Applied Spectroscopy Reviews
Biospectroscopy
Journal of Near-Infrared Spectroscopy
Spectrochimica Acta, Part A: Molecular and Biomolecular Spectroscopy
Vibrational Spectroscopy

Glossary of Terms

This section contains a glossary of terms, all of which are used in the text. It is not intended to be exhaustive, but to explain briefly those terms which often cause difficulties or may be confusing to the inexperienced reader.

Aetiology The medical study of the causation of disease.

Apodization A mathematical process used to remove the side lobes appearing in spectral bands, involving the multiplication of an interferogram by a suitable function before Fourier-transformation is carried out.

Attenuated total reflectance spectroscopy A reflectance infrared sampling method utilizing the phenomenon of total internal reflection.

Bending Molecular vibration involving a change in bond angle.

Chemometrics The application of statistical methods to the analysis of experimental chemical data.

Combination bands Infrared bands arising when fundamental bands absorbing at different frequencies absorb energy simultaneously.

Coupling Where a vibrational mode has contributions due to different molecular movements, which occurs when adjacent bonds have similar frequencies.

Deconvolution The process of compensating for the intrinsic linewidths of infrared bands in order to resolve overlapping bands.

Degeneracy Where different molecular vibrations occur with the same frequency.

Diffuse reflectance spectroscopy A reflectance infrared sampling method where the energy that penetrates one or more particles into a surface is reflected in all directions.

Infrared Spectroscopy: Fundamentals and Applications   B. Stuart
© 2004 John Wiley & Sons, Ltd   ISBNs: 0-470-85427-8 (HB); 0-470-85428-6 (PB)

Dispersive infrared spectrometer Spectrometer employing a dispersive element, such as a grating.

Doppler effect The effect in which radiation is shifted in frequency when the radiation source is moving towards or away from the observer.

Far-infrared region The region of the infrared spectrum occurring in the $400-100$ cm^{-1} range.

Fermi resonance When an overtone or a combination band has the same or similar frequency to a fundamental, bands appear split either side of the expected frequency value.

Fingerprint region The pattern appearing in the mid-infrared region resulting from skeletal vibrations.

Force constant The proportionality constant relating the stiffness of a molecular bond and the masses of the atoms at the ends of the bonds.

Fourier-transformation The mathematical method that converts data between the domains of distance and frequency.

Frequency The number of cycles per second of an electromagnetic wave.

Imaging An infrared technique where a large number of detector elements are read during the acquisition of spectra to produce a two- or three-dimensional picture of a sample.

Interferogram A signal produced as a function of the change of pathlength between the two beams in a Fourier-transform spectrometer.

Michelson interferometer The most common interferometer used in Fourier-transform spectrometry, consisting of two perpendicularly plane mirrors, one of which can travel in a direction perpendicular to the plane.

Mid-infrared region The region of the infrared spectrum occurring in the $4000-400$ cm^{-1} range.

Molar absorptivity The constant of proportionality relating the absorbance of a solution to the analyte concentration and the cell pathlength.

Near-infrared region The region of the infrared spectrum occurring in the $13\,000-4000$ cm^{-1} range.

Overtone bands Infrared bands arising from multiples of a fundamental absorption frequency.

Photoacoustic spectroscopy A reflectance sampling method, using intensity-modulated light absorbed by the surface of a sample located in an acoustically isolated chamber filled with an inert gas.

Selection rule A statement about which spectroscopic transitions are allowed for a particular process.

Skeletal vibrations Vibrations arising from the coupling of vibrations over a complete molecule.

Specular reflectance spectroscopy A reflectance infrared sampling method where the reflected angle of radiation equals the angle of incidence.

Stretching Vibration involving a change in bond length.

Vibration–rotation bands Bands due to the excitation of rotational motion during a vibrational transition.

Wavelength The distance between adjacent peaks of an electromagnetic wave.

Wavenumber The number of electromagnetic waves in a length of 1 cm.

SI Units and Physical Constants

SI Units

The SI system of units is generally used throughout this book. It should be noted, however, that according to present practice, there are some exceptions to this, for example, wavenumber (cm^{-1}) and ionization energy (eV).

Base SI units and physical quantities

Quantity	Symbol	SI unit	Symbol
length	l	metre	m
mass	m	kilogram	kg
time	t	second	s
electric current	I	ampere	A
thermodynamic temperature	T	kelvin	K
amount of substance	n	mole	mol
luminous intensity	I_v	candela	cd

Prefixes used for SI units

Factor	Prefix	Symbol
10^{21}	zetta	Z
10^{18}	exa	E

(continued overleaf)

Infrared Spectroscopy: Fundamentals and Applications   B. Stuart
© 2004 John Wiley & Sons, Ltd   ISBNs: 0-470-85427-8 (HB); 0-470-85428-6 (PB)

Infrared Spectroscopy: Fundamentals and Applications

Prefixes used for SI units *(continued)*

Factor	Prefix	Symbol
10^{15}	peta	P
10^{12}	tera	T
10^{9}	giga	G
10^{6}	mega	M
10^{3}	kilo	k
10^{2}	hecto	h
10	deca	da
10^{-1}	deci	d
10^{-2}	centi	c
10^{-3}	milli	m
10^{-6}	micro	μ
10^{-9}	nano	n
10^{-12}	pico	p
10^{-15}	femto	f
10^{-18}	atto	a
10^{-21}	zepto	z

Derived SI units with special names and symbols

Physical quantity	SI unit		Expression in terms of base or derived SI units
	Name	Symbol	
frequency	hertz	Hz	$1\ Hz = 1\ s^{-1}$
force	newton	N	$1\ N = 1\ kg\,m\,s^{-2}$
pressure; stress	pascal	Pa	$1\ Pa = 1\ N\,m^{-2}$
energy; work; quantity of heat	joule	J	$1\ J = 1\ Nm$
power	watt	W	$1\ W = 1\ J\,s^{-1}$
electric charge; quantity of electricity	coulomb	C	$1\ C = 1\ A\,s$
electric potential; potential difference; electromotive force; tension	volt	V	$1\ V = 1\ J\,C^{-1}$
electric capacitance	farad	F	$1\ F = 1\ C\,V^{-1}$
electric resistance	ohm	Ω	$1\ \Omega = 1\ V^{-1}$
electric conductance	siemens	S	$1\ S = 1\ \Omega^{-1}$
magnetic flux; flux of magnetic induction	weber	Wb	$1\ Wb = 1\ V\,s$
magnetic flux density; magnetic induction	tesla	T	$1\ T = 1\ Wb\,m^{-2}$

Derived SI units with special names and symbols *(continued)*

Physical quantity	SI unit		Expression in terms of base or derived SI units
	Name	Symbol	
inductance	henry	H	$1\ H = 1\ Wb\,A^{-1}$
Celsius temperature	degree Celsius	°C	$1°C = 1\ K$
luminous flux	lumen	lm	$1\ lm = 1\ cd\,sr$
illuminance	lux	lx	$1\ lx = 1\ lm\,m^{-2}$
activity (of a radionuclide)	becquerel	Bq	$1\ Bq = 1\ s^{-1}$
absorbed dose; specific energy	gray	Gy	$1\ Gy = 1\ J\,kg^{-1}$
dose equivalent	sievert	Sv	$1\ Sv = 1\ J\,kg^{-1}$
plane angle	radian	rad	1^{a}
solid angle	steradian	sr	1^{a}

[a] rad and sr may be included or omitted in expressions for the derived units.

Physical Constants

Recommended values of selected physical constants[a]

Constant	Symbol	Value
acceleration of free fall (acceleration due to gravity)	g_n	$9.806\ 65\ m\,s^{-2}$ [b]
atomic mass constant (unified atomic mass unit)	m_u	$1.660\ 540\ 2(10) \times 10^{-27}\ kg$
Avogadro constant	L, N_A	$6.022\ 136\ 7(36) \times 10^{23}\ mol^{-1}$
Boltzmann constant	k_B	$1.380\ 658(12) \times 10^{-23}\ J\,K^{-1}$
electron specific charge (charge-to-mass ratio)	$-e/m_e$	$-1.758\ 819 \times 10^{11}\ C\,kg^{-1}$
electron charge (elementary charge)	e	$1.602\ 177\ 33(49) \times 10^{-19}\ C$
Faraday constant	F	$9.648\ 530\ 9(29) \times 10^{4}\ C\,mol^{-1}$
ice-point temperature	T_{ice}	$273.15\ K$ [b]
molar gas constant	R	$8.314\ 510(70)\ J\,K^{-1}\ mol^{-1}$
molar volume of ideal gas (at 273.15 K and 101 325 Pa)	V_m	$22.414\ 10(19) \times 10^{-3}\ m^3\ mol^{-1}$
Planck constant	h	$6.626\ 075\ 5(40) \times 10^{-34}\ J\,s$
standard atmosphere	atm	$101\ 325\ Pa$ [b]
speed of light in vacuum	c	$2.997\ 924\ 58 \times 10^{8}\ m\,s^{-1}$ [b]

[a] Data are presented in their full precision, although often no more than the first four or five significant digits are used; figures in parentheses represent the standard deviation uncertainty in the least significant digits.
[b] Exactly defined values.

The Periodic Table

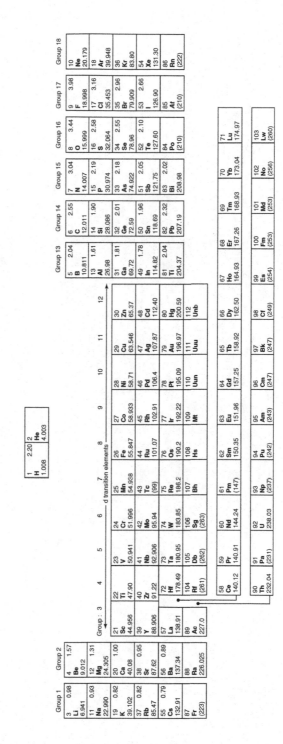

Index

Infrared Spectroscopy: Fundamentals and Applications B. Stuart
© 2004 John Wiley & Sons, Ltd ISBNs: 0-470-85427-8 (HB); 0-470-85428-6 (PB)

Carbon dioxide, 7, 10, 16, 42, 183, 188, 203
Carboxylic acids, 77, 79–80
Cells, 25–26
Cellulose, 179
Chemometrics, 67
Chloroform, 7–8, 50–51, 60–61, 92, 123
Cholesterol, 154–155, 161–162
Christiansen effect, 29
Cigarette smoke, 183–184, 203
Citronellal, 88–89
Classical least-squares method, 67
Clays, 108–109
Clinical chemistry, 161–162
Combination bands, 11, 178, 188, 194
Computers, 23, 49
Concentration, 57–59
Coordination compounds, 98, 102–104
Copolymers, 118–120, 198–199
Correlation table, 114–115
Cortisone acetate, 170–171
Coupling, 8, 12
Crystal field splitting, 126
Crystallinity, 126–127
Curve-fitting, 56–57, 143
Cyclohexane, 63–65

Deconvolution, 54–56, 143
Degeneracy, 7
Degradation, 132–135
Deoxyribonucleic acid, 49, 151–152, 155
Derivatives, 53–55, 143
Detectors, 16, 19–20
Deuterium oxide, 26–27, 141, 143, 146, 149–150
Deuterium triglycine sulfate detector, 19
Diamond anvil cell, 38, 190
Diatomic molecules, 5
Difference spectra, 52–53
Diffuse reflectance spectroscopy, 36–37, 131, 170, 179, 189
Dioctyl phthalate, 122
Dispersive infrared spectrometer, 16–18
Doppler effect, 6
Disease diagnosis, 152–155, 201
Drugs, 66–67, 168–174
Dyes, 180–182

Electromagnetic radiation, 2–5
Emission spectroscopy, 43
Environmental applications, 183–184
Ephedrine, 174, 202
Escherichia coli, 155–158
Esters, 77–78
Ethanol, 175–176
Ethers, 76–77, 79–80
Evolved gas analysis, 42–43

Far-infrared, 18, 20, 24, 48, 97
Fats, 174–176
Fellgett advantage, 23
Fermi resonance, 12, 78–79
Fibres, 121, 178–179
Films, 27–28, 31, 35
Fingerprint region, 11, 47
Focal plane array detectors, 39–40
Food science, 174–178
Force constant, 7–8
Fourier-transform infrared spectrometers, 18–25
Fourier-transformation, 18, 20–21
Frequency, 3
Fundamental bands, 8, 78, 188

Gallstones, 153–155
Gas chromatography–infrared spectroscopy, 41–42, 147–174, 183–184, 190
Gases, 31–32, 59, 183–184
Gaussian bandshapes, 55–57
Germanium, 34
Glucose, 161
Grains, 178–179
Gram–Schmidt chromatograms, 42–43
Grazing angle sampling, 35
Group frequencies, 46–48

Halogen-containing organic compounds, 82
Heisenberg Uncertainty Principle, 6
Heterocyclic compounds, 83
Hexanol, 58
Hierarchical clustering, 68, 158
Hydration, 97–98